ROUTLEDGE LIBRARY EDITIONS:
URBAN PLANNING

Volume 17

URBAN LAND AND PROPERTY MARKETS IN THE NETHERLANDS

URBAN LAND AND PROPERTY MARKETS IN THE NETHERLANDS

D. B. NEEDHAM, B. KRUIJT AND P. KOENDERS

Routledge
Taylor & Francis Group

LONDON AND NEW YORK

First published in 1993 by UCL Press

This edition first published in 2018
by Routledge
2 Park Square, Milton Park, Abingdon, Oxon OX14 4RN

and by Routledge
711 Third Avenue, New York, NY 10017

Routledge is an imprint of the Taylor & Francis Group, an informa business

British Library Cataloguing in Publication Data
A catalogue record for this book is available from the British Library

ISBN: 978-1-138-49611-8 (Set)
ISBN: 978-1-351-02214-9 (Set) (ebk)
ISBN: 978-1-138-48517-4 (Volume 17) (hbk)
ISBN: 978-1-138-48518-1 (Volume 17) (pbk)
ISBN: 978-1-351-05011-1 (Volume 17) (ebk)

Publisher's Note
The publisher has gone to great lengths to ensure the quality of this reprint but points out that some imperfections in the original copies may be apparent.

Disclaimer
The publisher has made every effort to trace copyright holders and would welcome correspondence from those they have been unable to trace.

Urban land and property markets in the Netherlands

D. B. Needham
Katholieke Universiteit Nijmegen

B. Kruijt
Universiteit van Amsterdam

P. Koenders
Katholieke Universiteit Nijmegen

UCL PRESS

First published in 1993 by UCL Press

UCL Press Limited
University College London
Gower Street
London WC1E 6BT

The name of University College London (UCL) is a registered
trade mark used by UCL Press with the consent of the owner.

ISBN:
1-85728-051-2 HB

A CIP catalogue record for this book
is available from the British Library.

Typeset in Times Roman.
Printed and bound by
Biddles Ltd, King's Lynn and Guildford, England.

CONTENTS

CONTENTS

PREFACE

This report was prepared under the supervision of Barrie Needham and Bert Kruijt. Most of the writing was done by Barrie Needham.

Most of the information was collected and sorted by Patrick Koenders. Information for most of the case studies was collected by Margot Verbiesen. Information for one of the case studies was collected by Eric Poell, for another case study by Dik Rouwenhorst, and for another by Tineke Kemp. The content of various sections was critically checked by Erwin van der Krabben. Valuable advice about some sections was given by Jos Janssen. Typing was by Mieke Verheggen and Annie van Bergen.

The research was carried out in the Department of Physical Planning (Vakgroep Planologie) of the University of Nijmegen, the Netherlands, where Barrie Needham is Senior Lecturer. Most of the others who helped with this project were temporary researchers or final-year students in that department. Bert Kruijt is Professor of Real Estate Economics at the University of Amsterdam.

The municipality of Nijmegen kindly provided help with preparing case study 6.1; the municipality of Amsterdam with the case studies 6.2, 6.3 and 10.2; the property advisers Jones Lang Wootton and the property developer Wilma Vastgoed with case study 6.3. Thanks are also due to the municipalities of Rotterdam (case study 10.1), Leeuwarden (case study 10.2), property developers Blauwhoed (case study 10.1) and MBO (case study 10.3).

The authors – Barrie Needham, Patrick Koenders, Bert Kruijt – wish to thank the other contributors for their help. The first version was produced in six months: it is clear that that would not have been possible if all those involved had not worked intensively and in close co-operation.

We wish to add two final points. First, at the beginning of September 1991, the value of the Dutch guilder (Hfl) was: £0.302, FF3.308, Lira 661.4, DM0.887, $0.508, ECU2.313. Secondly, in a work of this nature it is inevitable that some of the information it contains becomes out-of-date quite quickly. This applies in particular to the details of subsidy arrangements, conditions under which permits are granted, and such like. This book also contains many such details: the reader should remember this ageing process! But the book also outlines the way the land and property markets function in the Netherlands; and this information wears much better.

DBN   Nijmegen   OCTOBER 1992

ABBREVIATIONS AND ACRONYMS

AMRO   Amsterdam–Rotterdam Bank

AROB   Administratieve Rechtspraak Overheidsbeslissingen (Administrative Justice Act)

bn   billion

BTW   belasting op toegevoegde waarde (value added tax)

CBS   Centraal Bureau van de Statistiek (Central Bureau of Statistics)

CPB   Centraal Plan Bureau

DGVH   Directoraat Generaal Volkshuisvesting (General Directorate for Housing)

EC   European Community

EIA   environmental impact assessment

EIB   Economisch Instituut voor de Bouwnijverheid, Amsterdam (Economic Institute for the Building Industry)

ETI   Economic and Technical Institute

GNP   gross national product

ha   hectare

Hfl   Dutch guilder

IPR   Investeringspremie-regeling (regional development grant)

KADOR   Kadaster en Openbare Register (Cadastral Office)

M   million

MBO   maatschappij voor bedrijfsobjecten (a property development company)

Min. VROM   Ministerie Volkshuisvesting, Ruimtelijke Ordening en Milieu (Ministry of Housing, Physical Planning and Environment)

mld   milliard, 1,000 million

MPS   Meerjarenplan Stadsvernieuwing (multi-year plan for urban renewal)

NCIV   Nederlands Christelijk Instituut voor de Volkhuisvesting (National Protestant Association of Housing Associations)

NEPROM   Vereniging van Nederlandse Projectontwikkeling Maatschappijen (National Association of Property Developers)

NIROV   Nederlands Instituut voor Ruimtelijke Ordening en Volkhuisvesting (Dutch Association for Planning and Housing)

NMP   National Plan for the Environment
NRO   Nationaal Rayon Onderzoek (National Survey of Housing Market Regions)
NVM   Nederlandse Vereniging van Makelaars in Onroerende Goederen (Dutch Association of Estate Agents)
NWR   Nationale Woningraad (National Association of Housing Associations)
OTB   Onderzoeksinstituut Technische Bedrijfskunde
RAVI   Raad voor Vastgoedinformatie (Advisory Council for Real Estate Information)
RIGO   Research Institute for the Built Environment
ROM   Ruimtelijke Ordening en Milieubeheer (a journal)
SCP   Sociaal en Cultureel Planbureau
SEO   Stichting voor Economisch Onderzoek (Foundation for Economic Research)
TAUW   name of a firm
VAT   value added tax (see BTW)
VGM   VastGoedMarkt (a journal title)
VVG   Vereniging van Grondbedrijven (Association of Land Departments)
WBO   Woningbehoefte-onderzoek (Housing Needs Survey)
WRR   Wetenschappelijke Raad voor het Regeringsbeleid (Netherlands Scientific Council for Government Policy)

PART I
Overview

Groningen

NOORD-NEDERLAND

* 2
* 1

Friesland

* 3

Drenthe

Noord-Holland

* 9

* 4

Overijssel

14 13
*   *

10
*

* 6   OOST-NEDERLAND

* 5

WEST-NEDERLAND

12 * *
       11

* 15

Utrecht

Gelderland
* 7

Zuid-Holland
* 16

* 8

Noord-Brabant   * 20

Zeeland
* 17

* 18   * 19   ZUID-NEDERLAND

* 21

Limburg

23 *

* 22

0   20   40   60   80 km

1 = Groningen
2 = Ieeuwarden
3 = Assen
4 = Zwolle
5 = Enschede
6 = Apeldoorn
7 = Arnhem
8 = Nijmegen
9 = Lelystad
10 = Almere
11 = Amersfoort
12 = Utrecht
13 = Amsterdam
14 = Haarlem
15 = 's-Gravenhage
      (Den Haag)
16 = Rotterdam
17 = Middelburg
18 = Breda
19 = Tilburg
20 = 's-Hertogenbosch
      (Den Bosch)
21 = Eindhoven
22 = Maastricht
23 = Heerlen

CHAPTER 1

Basic information

1.1 The constitutional and legal framework

The Kingdom of the Netherlands calls itself a decentralized, unitary state, with a system of constitutional parliamentary monarchy. "Decentralized" means that there are governmental bodies (bodies with the public authority to govern) for not only the whole nation but also every province, and every municipality (*gemeente*). Some other bodies, with non-territorial responsibilities, also have governmental powers. "Unitary" means that governmental bodies must not work against each other. This is realized constitutionally by the national government allowing provincial and municipal governments to act only in ways specified by legislation passed nationally, and by making the provincial governments subordinate to the national government, and the municipalities subordinate to both provincial and national governments.

Governmental bodies with territorial responsibilities are of three types: national government, the governments of the 12 provinces and municipal governments (there are now fewer than 700 municipalities, a number gradually being reduced by amalgamations). Governmental bodies with functional rather than territorial responsibilities are the water boards (*waterschappen*), corporations for industry and the professions (*publiekrechtelijke bedrijfsorganisaties*), and some others (*andere openbare lichamen*).

All of these public authorities (governmental bodies) have powers derived from the constitution. This gives them a certain autonomy. But it is limited, not only by the substantive specification of the limits but also by the supervision that a higher authority can exercise over a lower. A higher authority can use supervision to ensure that a lower authority does nothing inconsistent with the policy being pursued by the higher. In practice, therefore, autonomy is severely limited and much of what a lower authority does falls under the description joint-rule (*medebewind*). The actions of a lower authority with respect to a certain policy field are therefore taken partly to execute the policy of a higher authority, although that policy has been formulated so as to allow latitude for interpretation. A lower authority

3

cannot choose whether or not to execute a higher authority's policy, but it is free to interpret it within specified limits. There is talk also of complementary government (*complementair bestuur*), when different public authorities voluntarily pool their autonomous powers to pursue a certain policy: but in practice this does not happen often.

These principles governing the relationship between the three territorial authorities (national, provincial, municipal) make it difficult to describe the division of tasks between them. A higher authority may decide to undertake certain activities with which it had not previously been involved, thus precluding a lower authority from so doing. Nevertheless, it is helpful to say something general about the division of tasks in practice. A good indication is given by the relative expenditure of the three levels, as indicated in Table 1.1.

Table 1.1 Expenditure of national, provincial and municipal government.

(mln Hfl)	National (1988)	Provincial (1988)	Municipal (1990)
personnel costs	21,262	1,116	12,995
transfer payments (social benefits etc)	83,139	2,871	22,833

Source: national and provincial; Central Bureau of Statistics (CBS), Statistisch Zakboek 1991, municipal; Central Bureau of Statistics (CBS), Statistiek der Gemeentebegroting 1990.

Table 1.1 makes it obvious that the provinces are not "spending departments". Their task is mainly co-ordination and supervision of the municipalities which are, equally clearly, heavy spenders.

How the provinces and municipalities spend their money is indicated in Tables 1.2 and 1.3, which include both current and capital expenditure, and the grants and transfers (current and capital) made to others.

The heavy expenditure by municipalities on social services (current), and on planning and housing (capital), is caused largely by their passing on grants made available by central government. The division of powers and responsibilities between national, provincial and municipal governments with respect to land and property is described elsewhere.

Public administration

The head of state is the monarch, who as such is inviolable. Monarch and ministers together constitute the government or Crown. The ministers and secretaries of state form the cabinet. When in conference, the ministers form the council of ministers, headed by the minister-president. Parliament

4

Table 1.2 Expenditure by municipalities, 1990.

(mln Hfl)	Current	Capital
general purposes	4,457	2,477
law and order	2,710	660
traffic, transport, water	3,437	5,441
economic affairs	363	3,225
education	5,880	8,239
culture and recreation	5,224	3,964
social services	16,779	324
public health	2,895	5,393
physical planning and housing	8,519	55,763

Source: Current, CBS, Statistisch Zakboek 1991; Capital, CBS,
Statistiek der Gemeentebegroting, 1990. Both exclude those municipal
agencies which have their own accounts.

Table 1.3 Expenditure by provinces, 1988.

(mln Hfl)	Current	Capital
general purposes	261	87
law and order	14	1
traffic and transport	1,718	389
water management	118	19
environmental protection	755	190
recreation	77	18
economic and agricultural affairs	405	228
welfare	2,888	21
physical planning and housing	293	85

Source: CBS, Statistisch Zakboek 1991.

(*Staten-generaal*) consists of a first and a second chamber. The second chamber is made up of 150 representatives chosen directly by the Dutch electorate. Members of the second chamber have more powers than their 75 indirectly elected counterparts in the first chamber, which acts more or less as a controlling body in the last resort. Ministers are answerable to parliament without being members of it (Gilhuis 1991).

The formal powers of a municipality are divided between three governing bodies: the mayor (*burgemeester*), the municipal executive (council of mayor and aldermen), and the council (*gemeenteraad*). Mayors are appointed by the Crown, after private consultations with local politicians; the mayor of a big city holds one of the most important public offices in the country. Councils are elected every four years by local electorates. Each council in turn chooses between three and six of its members to be aldermen: the number depends on the municipality's size, which also determines whether they are full- or part-time. Unlike councillors, they receive a salary. Chapters 3 and 7 describe how formal powers with respect to land and property markets are divided between mayor, municipal executive, and council.

The administrative structure of a province is similar. A Queen's commissioner (*commissaris der koningin*) is appointed by the Crown, and holds one of the top public offices. The provincial council (*provinciale staten*) is composed of elected politicians who choose the deputies-general (*gedeputeerden*). These deputies and the Queen's commissioner constitute the provincial executive (*gedeputeerde staten*). Each of the bodies (commissioner, executive, council) has its own powers and responsibilities.

It is important to note that neither municipalities nor provinces raise more than a small part of their income from local taxes: both depend heavily on grants from national government. These grants have two forms: general and specific. A general grant can be spent as a public authority sees fit (but always within its delimited "policy space" discussed above). A specific grant has to be spent executing the policy for which it was granted: a specific grant is appropriate when a lower authority executes central government policy under the "joint-rule" arrangement.

The sources of current income for municipalities (1985) and their relative importance are given in Table 1.4.

Table 1.4 Sources of current income of municipalities, 1985, in percentages.

local taxes (mainly the property tax)	8.2
central grants	25.8
specific grants	60.8
income from municipal property (rents etc)	5.2
Total	100.0

Source: Koopmans, 1987, p. 125.

Note: the property tax is described in more detail in Ch.3.

It is clear that the formal autonomy that a lower authority legally enjoys can be restricted if it receives much of its income from a higher authority, which uses its power over funds to govern by the purse strings. This danger is always present in the Netherlands. The financially weak position of the municipality, with such a large part of current income in the form of earmarked central government grants, and its constitutionally weak position (its autonomy is specified very loosely), might suggest that public administration is very centralized. This is not so. In practice, working arrangements have grown up, and are maintained, by which municipalities and, to a lesser extent, provinces have a great deal of autonomy. This is particularly so where physical planning policy is concerned.

The legislative process

Legislative power lies with the government and parliament. Together, they constitute the formal legislator and both can take the initiative in proposing legislation. For a bill to become law it must be approved by both the second chamber and the first chamber.

The act thus passed (*wet*) is called formal legislation. This may, within certain limits, transfer authority to issue decrees and regulations to other governing bodies (e.g. the government, ministers, provinces, municipalities, water authorities, and corporations for industry and the professions). Such decrees and regulations are called "legislation in a material sense". When issued by government by virtue of an act of parliament, they are known as general administrative orders (*algemene maatregelen van bestuur*).

More often than not, formal legislation has the character of framework legislation. This means that the formal legislation says very little about the content of the regulations (such as the values of the standards to be applied). The importance of the formal law is that it sets the framework and then delegates to government, provinces and municipalities the power to specify the content of the policy by filling in the framework. This is important in the context of land and property law discussed in Chapters 3 and 7.

Formal legislation is one of the four sources of law in the Netherlands. The others are common law (which is only supplementary to formal legislation), international treaties (which have priority over national legislation), and case law (which is binding on other cases only if the supreme court decides so explicitly). For a fuller account of the legislative process, see Gilhuis (1991) and Aalders (1991).

Rights of ownership over land and property

Rights of ownership over land and property are regulated in principle by the civil code (*Nieuw Burgerlijk Wetboek*), the new version of which came into force on 1 January 1992. All references here are to this new version. Book 5, Article 1, says: "Ownership is the most comprehensive right that a person can have over something. The owner is free to use that thing, and to exclude others from using it, as long as that use does not infringe the rights of others, nor transgress limitations arising out of written or unwritten law. The owner of the thing is owner also of the products thereof, unless others have acquired a right to those products." This principle relates to all property, movable and immovable: it gives the owner of landed property the same

7

general rights and obligations as the owner of a motor car, a painting, or a fountain pen! As a consequence, there is little legislation outside the civil code relating specifically to land and buildings.

The law relating to civil contracts (the legislation that applies to all legal persons, including public agencies, when they enter into voluntary contracts) is built around that principle. Public law relating to landed property (by which public authorities may impose obligations on private persons and agencies) has to take account of that same principle.

The civil code (Book 5) qualifies those rights of ownership separately for movable and immovable property. Title 3 regulates the general principles relating specifically to landed (immovable) property. An important feature is that the physical boundaries of a unit of landed property must be established by an act of a public notary and entered in a public register within 14 days (Ch. 3.1). Title 4 regulates the rights and obligations of the owners of adjacent property rights, such as that the owner of a landed property may not cause nuisance to the owners of other properties. Title 6 regulates easements or servitudes, with which the rights of ownership may be encumbered. These consist of an obligation to allow something to take place on another property and to desist from doing something on one's own property. Such encumbrances must be legally specified in particular cases. Title 7 regulates ground leases, Title 8 the right to have and use building works or planting on land owned by another, Title 9 condominium rights.

In another part of the civil code (Book 3, Article 89) it is laid down that the title to property can be transferred only by an act of a public notary, after which the change must be entered in a public register.

When the right of ownership over a unit of landed property is transferred, the existing owner may include in the contract both restrictive covenants and positive obligations that are binding on the new owner. Moreover, the existing owner may include conditions that are binding not only on the first new owner, but on all subsequent ones. These are called chain agreements, and are regulated by the civil code (Book 6, Article 252), which limits them to restrictive covenants and to permitting another to do things that affect your ownership rights. The way municipalities make use of such conditions when disposing of land is discussed in Chapter 3.1.

In these ways, the civil code restricts ownership rights in general, for instance in respect of nuisance to adjacent properties, allows them to be restricted in general by public law (such as the law on physical planning), and allows them to be restricted in particular cases by private contracts (e.g. by encumbrances). In the latter case, the legal owner of landed property grants to others specific rights over his property, thus restricting his own rights of ownership, and possibly also the rights of subsequent owners. When the general nature of those rights thus granted are specified in the civil code, they have the form of a real or material right (*zakelijk recht*: rights *in rem*): they are attached to the property not to the owner, and change of owner does not change such real rights. To recapitulate, these rights are ownership, servitudes, ground leases, building rights and condominium rights.

The owner of property may grant to others another type of right over his property, called a contractual right (*persoonlijk recht*: rights *in persona*). This right is valid only between the parties named in the contract: if one of those parties withdraws, for instance because of death, or if the legal owner of the property sells his rights, the contractual right expires. The most important example of a contractual right is a tenancy to rent. It follows therefore that a tenancy cannot be transferred.

When the legal owner of a unit of landed property grants to another rights that are particularly extensive and durable, the enjoyer of them is regarded legally as having economic ownership of the property (as distinct from legal ownership). This is a recent development; it is not based in civil law but has grown out of taxation law and was legally recognized for the first time in 1955. The person who has economic ownership has the complete economic interest in the property (including the risk associated with changes in value). Also, the owner of the economic rights has to include the property as an asset on the balance sheet; and he may write off the value of the ownership against tax. The legal title remains with the vendor for a period to be agreed upon, after which it is transferred to the economic owner. If, in the meantime, the economic owner sells that right, the legal title may be transferred to the new economic owner (but he has a weaker right to it than the first economic owner). Rights that are less extensive and durable (e.g. rights of tenancy) are regarded as user rights, not ownership rights.

All the rights so far described fall under private (civil) law: two parties enter voluntarily into an agreement; if they later disagree, that must be resolved in the civil courts. However, such private contracts are increasingly coming under the influence of public law, in particular the criminal code. An important example is Article 461 concerning trespass, which is being used to enforce certain civil contracts.

The civil code regulates inheritance as well (Book 4). For inheritance purposes, landed property is treated just as any other property. There is no special regulation to prevent farmland being divided into small and inefficient units, as can be found in some countries. For discussion of death duties, see Chapter 7.3.

The ways in which public law can restrict rights of ownership over landed property, allowing public authorities to impose restrictions on that right without any voluntary agreement, are described in Chapters 3.1 and 7.1.

1.2 The economic framework

Economic variables and how they have changed
Table 1.5 presents key economic variables for selected years since 1970.

Table 1.5 Key economic variables.

	1970[1]	1975[1]	1980	1985	1987	1989	1990[2]	1991[2]
Gross national product								
absolute (bln.Hfl)	121,740	219,940	336,120	418,860	430,500	474,150	503,800	530,350
per head (Hfl)[3]	9,395	16,173	23,854	28,981	29,456	32,026	33,827	35,335
The economic sectors								
(% of GNP)[3,5]								
primary[10]		9.0	9.8	13.4	8.3	7.9		
secondary		32.8	29.0	26.8	28.2	29.8		
tertiary		58.2	61.2	59.8	63.5	62.3		
Investment (fixed capital formation, gross)								
absolute (bln. Hfl)								
residential buildings		11,690	20,840	20,380	22,040	26,150		
non-residential buildings	15,120	12,480 8,70	14,250	18,820				
civil engineering works		6,470	8,030	8,030	7,510	8,140		
external transport equipment		5,100	7,080	10,360	11,310	13,020		
increase in livestock		(20)	60	(270)	(540)	20		
other machinery and equipment		13,220	17,660	27,290	29,560	34,550		
transfer costs on purchase of used fixed assets		1,160	2,000	1,980	2,560	2,850		
Total fixed capital formation (gross)		46,320	70,790	80,250	86,690	103,550		
% of GNP[3]		21.1	21.1	19.2	20.1	21.8		
Level of income[9]								
absolute (bln. Hfl)	88.75	139.22	212.58	271.44	282.93	318.55	344.30	360.35
per head (Hfl)[3]	6,849	10,237	15,086	18,780	19,359	21,516	23,118	24,009
Cost of living (1980=100)		74.4	100	122.3	·122.0	124.4	127.2	
Savings								
absolute (bln. Hfl)		50,870	67,610	99,540	90,520	115,070		
% of Gross National Income[3]		23.3	20.3	24.0	21.3	24.5		

Table 1.5 continued

	1970[1]	1975[1]	1980	1985	1987	1989	1990[2]	1991[2]
Balance of payments (current account)[4]								
exports (bln. Hfl)			204,040	305,660	262,270	328,920		
imports (bln.Hfl)			208,940	288,350	256,140	313,150		
surplus of the nation on current transactions			(4,900)	17,310	6,130	15,770	16,000	17000
Capital mobility[4] (capital account)								
exports (bln. Hfl)			(4,230)	18,420	7,150	16,990		
imports (bln. Hfl)			1,650	1,720	2,590	2,920		
change in financial assets	5,280	(5,880)	16,700	4,560	14,070			
Interest rates (%) on newly registered first normal mortgages on real estate	9.33	10.15	7.77	6.99	7.58	8.7		
redeemable bonds (long-term interest)[6]	8.42	10.11	7.25	6.32	7.24	8.9		8.8
De Nederlandsche Bank-discount on bills of exchange		4.5	8.0	5.0	3.75	7.0		
Unemployment as % of the population of working age[7]		4.9	7.4	16.0	14.2	13.2	12.7	
registered unemployment (×1000)		260	325	761	685	390	365	355
registered as seeking work (×1000)[8]						660	625	610
Registered as unfit for work as % of the population of working age			14.6	16.1	16.4	16.9	17.7	
absolute (×1000)			661	768	793	847	870	

Sources: CBS, Central Plan Bureau (CPB), OECD, Sociaal Cultureel Planbureau (SCP).

(1) Figures for 1970 and 1975 are given if they are available and illuminating. (2) Figures for 1990 are often not yet available. The figures for 1991 are estimates made by the Central Bureau of Statistics (CBS) or the Central Plan Bureau. (3) Our own calculations. (4) The way of collecting these figures was changed in 1985. We have made a conversion. (5) Figures for the different types of industry per sector are given in I.1.5.1. (6) This represents the interest rate on a secure investment. (7) The unemployment rate for those born abroad was in 1987–88: from Turkey 44%; from Morocco 42%; from Surinam 27% from the Antilles 23%; from the Southern European EC member states 18%; from other countries except EC member states 36%. (8) This distinction was introduced in 1987. The "registered unemployed" are those receiving unemployment benefit, but there are in addition others who are seeking work. (9) Gross final income plus transfer payments and pensions (CPB, 1990). (10) The figures (they represent value added) for 1985 are correct but extra-ordinary: in that year there was an unusually large export of natural gas. Generally because of a change in definitions, many statistics cannot be given in a comparable way for 1981, 1982, 1983. In many cases, a continuous change can be interpolated for those years. However, it is possible that there was a fall in some economic variables (especially investment) because of the economic recession.

Changes in the volume and structure of employment are presented in Table 1.6. The decline, both absolute and relative, in the primary and secondary sectors is very clear.

Table 1.6 Changes in the volume and structure of employment.

Industrial sector	1975		1980		1985		1987	
	abs	%	abs	%	abs	%	abs	%
0	299	6.4	278	5.8	271	5.9	267	5.6
1	6	0.1	8	0.2	9	0.2	10	0.2
2/3	1112	23.8	993	20.7	890	19.4	912	19.2
4	44	0.9	46	1.0	46	1.0	46	1.0
5	429	9.2	449	9.3	330	7.2	349	7.4
6	903	19.3	939	19.5	884	19.2	925	19.5
7	309	6.6	320	6.7	320	7.0	335	7.1
8	443	9.5	508	10.6	527	11.5	549	11.6
9	1125	24.1	1266	26.3	1318	28.7	1348	28.4
total	4670	100	4807	100	4598	100	4741	100

Source: CBS

Note the absolute figures are in 1000 man years.

The industrial sectors are: 0 agriculture and fishing; 1 mineral extraction; 2/3 manufacturing; 4 public utilities; 5 construction; 6 retail and wholesale trade, cafés and restaurants, repairs; 7 transport, storage and communication; 8 banking, insurance, commercial services; 9 other services, including public administration.

At the same time as employment has been changing, so has the number of people seeking work. The "balancing item" is unemployment. This is illustrated in Table 1.7.

Some striking changes have taken place in the working population. The large increase in unemployment after 1977 was caused both by economic and demographic factors. The recession at the beginning of the 1980s caused the number of jobs to fall, while at the same time many young people born during the baby boom came into the labour market for the first time, substantially increasing unemployment (Sec. 1.3). The fall in unemployment at the end of the 1980s is caused partly by a broader interpretation of "unfit to work". More people, who would previously have drawn unemployment benefit, were classified in this way.

Definitional problems make it difficult to specify the relationship between working time and leisure time and how this has changed, but the number of days paid leave per year is given in Table 1.8 (for 1986).

Prognoses

The Central Plan Bureau (CPB) regularly publishes prognoses of how the Dutch economy might change. The following are taken from the 1990 publication *Economisch Beeld 1991 (The economic situation in 1991)*. For the next few years, economic development will be positive. Exports will grow by 6.5% a year, partly because of changes in unified Germany, partly because of increased investment, and partly because prices will become more competitive. If imports remain the same, in 1991 there should be a surplus of Hfl 17 billion on the balance of trade current account.

Table 1.7 Employment and unemployment.

	Employees (M + F)	Registered unemployed[1] (M + F)	Employees + unemployed (M + F)
1979 compared with 1974[2]			
absolute 1979	3,902,300	209,700(5.1%)	4,112,000
absolute change	205,000	77,000	282,000
change as % 1974	5.2	58.0	7.0
1981 compared with 1979			
absolute 1981	3,827,000	343,880(8.2%)	4,171,500
absolute change	(74,500)	134,100	59,500
change as % 1979	(1.8)	63.9	1.4
1983 compared with 1981			
absolute 1983	3,913,400	767,700(16.4%)	4,681,100
absolute change	85,600	423,900	509,600
change as % 1981	2.2	123.3	12.2
1985 compared with 1983			
absolute 1985	4,087,500	778,500(16.0%)	4,866,000
absolute change	174,100	10,800	184,900
change as % 1983	4.5	1.4	3.9
1987 compared with 1985			
absolute 1987	4,187,300	693,000(14.2%)	4,880,300
absolute change	99,800	(85,500)	14,300
change as % 1985	2.4	(11.0)	0.3
1988 compared with 1987			
absolute 1988	4,203,600	684,300(14.0%)	4,887,900
absolute change	16,300	(8,700)	7,600
change as % 1987	0.4	(1.3)	0.2

Source: CBS, Sociale Maandstatistiek.

Notes: (1) Between brackets, as a percentage of the population of working age.

(2) All figures give the situation on 1 March.

Gross national product is expected to grow in the next few years by about 3% a year in real terms, which is higher than in the other European Community (EC) countries except Germany. Growth will be faster in manufacturing (especially bulk chemicals) and in services (especially trade and transport), and slower in construction, with stabilization of investment

Table 1.8 Holiday entitlements.

	Holidays and feast days	ADV-days[1]
in public sector	31.6	9.2
in subsidized sector	24.5	11.2
in private sector	24.6	7.3

Source: Beckers and v.d. Poel, 1989, p. 130.

Note: (1) An ADV-day is a day's leave given in lieu of a wage rise; the aim was to redistribute employment.

in dwellings and industrial space, and in agriculture (affected by EC policy and environmental policy) with the exception of horticulture.

Employment is expected to continue to grow by about 1% a year, with the fastest growth in the tertiary sector. The growth will be faster than the growth in labour supply, so the number of unemployed should decline slowly (in 1990 and 1991 by 15,000, to around 510,000). Inflation will be low. The high level of interest rates is a disturbing factor. It is caused mainly by developments in Germany and is damaging for housing construction and government finance.

The level of public expenditure on social benefits is another disturbance to the economy. It is not that the income to pay for these is falling, but that expenditure is increasing greatly. An important cause of this is the increasing number of people on sick leave and registered as unfit to work. Nobody can explain this and it was a political hot potato in 1991.

1.3 The social framework

Demographic changes

Table 1.9 summarizes the most important demographic changes since 1980 and gives the results of some forecasts to 2000.

The structure of the population by household and age is given in Tables 1.10 and 1.11 (Table 1.10 refers to 1987). A growing part of this total population is made up of non-Dutch nationals. Tables 1.12 and 1.13 give some figures on this. The distribution of the total population between the 12 provinces (1990) is shown in Table 1.14. Table 1.15 shows the distribution between different types of municipality.

We see that the proportion of the total population living in urban municipalities is falling, and that living in urbanized rural municipalities is rising.

The population composition is not uniform throughout the country.

Table 1.9 Demographic data and forecasts.

	1980[1]	1982	1984	1986	1988	1990	1995[2]	2000[2]
Population (×1,000,000)	14.1	14.3	14.5	14.6	14.8	15.0	15.3	15.7
Population per km²,of land[3]	419	423	426	431	436	441	450	462
Population by age[4]								
0–19 yrs	30.9	29.7	28.3	27.1	26.2	18.2	18.2	18.3
20–64 yrs	57.5	59.2	59.8	60.5	61.2	68.9	68.4	67.9
65 and older	11.6	11.8	12.0	12.4	12.7	12.9	13.4	13.8
Live births (×1,000)	181.3	170.2	178.1	184.5	187.0			
Deaths (×1,000)	114.3	118.0	123.0	122.2	124.2			
Immigration (×1,000)	112.5	67.0	79.0	96.0	91.0			
Emigration (×1,000)	59.5	61.0	55.0	52.0	56.0			
Average annual increase (%) since preceding year	0.83	0.38	0.41	0.59	0.61			
Life expectancy at birth, males	72.5			73.1	73.7			
females	79.2			79.6	80.2			
Average age at death								
males	69.2			70.3	70.5			
females	74.2			76.0	76.6			
Households: (×1,000)								
multi-person households	3,921	3,999	4,067	4,107	4,222			
single person households	1,085	1,132	1,353	1,604	1,713			

Source: CBS, Statistisch Zakboek.

(1) All dates refer to 31 December.

(2) Source for these two columns: DGVH/DOK/SEO-prognose's. In: Min. VROM, 1990(c), p. 51

(3) Here "land" is meant literally = 34,000 km². The total surface area of the Netherlands, inclusive of water is 41,864 km².

(4) These figures are per 100 inhabitants

Up to 19 years; non-productive (school attenders). From 1990 the youngest group is up to 15 years (school attenders). Productive from 19 (from 1990, from 15) to 65 years. After 65 years; pensionable age.

Table 1.10 The structure of population by households.

x 1,000	Living in nuclear family			Not living in nuclear family		Total	
	husband/ wife	single parent	child	males	females	males	females
<15 yrs	-	0	2,720	13	11	1,403	1,342
15–24 yrs	215	17	1,772	225	267	1,274	1,221
25–34 yrs	1,465	80	204	388	266	1,225	1,179
35–44 yrs	1,752	119	49	182	94	1,129	1,067
45–54 yrs	1,280	82	22	111	83	802	775
55–64 yrs	1,076	54	8	93	161	669	724
65–74 yrs	673	31	1	82	263	459	591
75 and older	263	27	0	97	367	263	492
Total	6,725	409	4,775	1,193	1,512	7,223	7,390

Source: CBS.

Table 1.11 The number of nuclear families by number of family members.

x 1,000	Number of nuclear families						Non-family persons
	2 persons	3 persons	4 persons	5 persons	6 or more persons	total	
31 May 1960	830	590	553	329	415	2,718	1,163
28 February 1971	1,058	713	751	395	346	3,261	1,439
1 January 1987	1,473	836	1,023	328	111	3,771	2,705

Source: CBS.

Table 1.12 Numbers of non-Dutch nationals.

	1980	1985	1987	1988	1989
Non-Dutch nationals					
% of the total population	3.4	3.9	3.9	4.0	4.2
absolute total (x 1,000)	520.9	558.7	568.0	591.8	623.7[1]

Source: CBS

Note: (1) The most important countries of origin in 1989 are: Turkey (176,500), Morocco (139,200), Federal Republic of Germany (40,300), United Kingdom (37,400).

Table 1.14 The distribution of population between the provinces, 1990.

	Total population (x 1,000)	Inhabitants per m²
Groningen	553.8	235
Friesland	599.1	178
Drenthe	442.2	166
Overijssel	1,020.4	306
Flevoland	212.5	150
Gelderland	1,804.0	360
Utrecht	1,015.5	745
North-Holland	2,375.9	893
South-Holland	3,219.8	1,120
Zeeland	355.9	198
North-Brabant	2,189.4	443
Limburg	1,103.9	509
Netherlands	14,891.9	439

Source : CBS.

ble 1.13 The population of non-Dutch nationals by e.

	Absolute	Per 100 non-Dutch nationals	Per 100 inhabitants
19 yrs	220,990	35.4	1.5
–59 yrs	379,169	60.8	2.6
and older	23,510	3.8	0.2
tal	623,769	100.0	4.2

rce: CBS

Table 1.15 The distribution of the population by type of municipality.

	1980 a	1980 b	1980 c	1989 a	1989 b	1989 c
Rural municipalities[1]	239	1,639.0	11.6	196.0	1,688.7	11.4
Urbanised rural municipalities of which:	452	5,116.3	36.3	389.0	5,593.9	37.8
industrialised rural municipalities[2]	326	3,128.1	22.2	274.0	3,411.1	23.0
specific commuter municipalities[3]	126	1,988.2	14.1	115.0	2,182.8	14.7
Urban municipalities[4] of which:	120	7,334.2	52.0	117.0	7,521.1	50.8
with 100,000 or more inhabitants	15	3,567.9	25.3	15.0	3,542.5	23.9
total	811	14,089.5	100	702	14,803.8	100

a = number of municipalities, b = population x 1,000, c = population in % of total population (people on the central population register are excluded)

Source: CBS.

Notes: (1) Municipalities where over 20% of the male labour force are employed in agriculture.

(2) Municipalities with less 20% male labour force employed in agriculture, whose largest residential nucleus counts fewer than 30,000 inhabitants and which does not have an obvious function as a regional centre.

(3) Urban municipalities, garden suburbs, commuter villages etc. with less than 20% agricultural population and at least 30% resident commuters, of whom over 60% do not originate from the municipality concerned.

(4) Small (10,000–30,000 inhabitants), medium-sized (30,000–100,000) and large (over 100,000) towns and cities. Dordrecht and Leiden are not counted as large cities as the residential nucleus on 28 February 1971 counted fewer than 100,000 inhabitants.

Variations between big cities, regions, and different sorts of municipality are shown in Tables 1.16 and 1.17.

The large increase in the proportion of one-person households is noticeable, as is the fact that this proportion is higher in the cities. The differences between the regions are not great.

More detail about how household structure varies between the big cities of the Randstad and the small authorities elsewhere is shown in Table 1.18.

Some of these demographic changes could have important effects on the market for urban land.

The increase in the percentage of people aged over 65 has consequences for not only the social security system but also housing (the number of dwellings, type and location, and access to facilities). Predictions indicate 24% of the population will be older than 64 in 2035 (Uitterhoeve 1990: 38). Note that life expectancy is also increasing.

Table 1.16 The distribution of households by size.

| | Large cities | | Netherlands | |
| | 1978 | 1988 | 1978 | 1988 |
	%	%	%	%
1 person	19	43	10	27
2 persons	32	30	26	30
3 persons	19	13	19	16
4 persons	20	11	28	19
5 persons	5	3	11	6
6 and more persons	5	1	6	2
total	100	100	100	100

Source : NRO, unweighted.

Table 1.17 The distribution of households by size and location.

	North[1]	East[2]	South[3]	West[4]	Cities	Netherl.
1 person	27	23	22	24	43	27
2 persons	31	30	29	31	30	30
3 persons	14	14	19	16	13	16
4 persons	19	22	22	21	11	19
5 persons	7	8	7	6	3	6
6 or more persons	2	3	2	2	1	2
total	100	100	100	100	100	100

Source: NRO.

(1) includes the provinces Friesland, Groningen and Drenthe.

(2) includes the provinces Overijssel, Gelderland and Flevoland.

(3) includes the provinces Zeeland, North-Brabant and Limburg.

(4) includes the provinces Utrecht, North-Holland and South-Holland.

Table 1.18 Variations in household structure between the big cities of the Randstad.

	1977 %	1985 %
Single person households as a % of all households:		
large local authorities in the Randstad	12	37
small local authorities outside the Randstad	5	18
Families with children as a % of all households:		
large local authorities in the Randstad	48	27
small local authorities outside the Randstad	63	48
single person households aged 18–34 years *in % of all single person households:*		
large local authorities in the Randstad	15	37
small local authorities outside the Randstad	13	25

Source : SCP.

Reduction in the size of the household, associated with an increase in the number of small households (one and two persons), has significant consequences for the housing market: more separate dwellings, an increase in land use, and a more differentiated housing market. In the cities this development is clearly visible.

Also apparent is the concentration of ethnic minorities in the cities: their share of the Dutch population is increasing slightly and is about 5%. In Amsterdam they make up 20% of the population (WRR 1990: 93).

There are also differences in the age structure of the population in towns and cities, with the cities' populations "growing older". However, the steady reduction in the number of pupils at primary schools in the cities has now ended, partly because of the influx of ethnic minorities (WRR 1990: 139).

Mobility

Table 1.19 gives figures for housing mobility, which is measured in terms of turnover rate, defined as "the number of dwellings of a certain type where a removal has taken place in a given year expressed as a percentage of the total dwellings of that type in that year" (Min. VROM 1990c).

From Table 1.19 we can conclude that: for both tenants and owners, mobility has decreased; for tenants, mobility is greater the more expensive the dwelling; owners move about half as often as tenants. Table 1.20 shows the sectors from which movers came and the sectors to which they went. The category "none" refers to first-timers in the housing market.

Table 1.20 indicates that more people moved from rented accommodation to bought than from bought to rented, and that most first-time buyers start in the rented sector. In 10 years there has not been much change in this pattern. Another aspect of mobility is movement from one part of the country to another. Table 1.21 shows net internal migration (i.e. ignoring movements across national boundaries) between regions, the annual average between 1980 and 1987 and the figures for 1988. (See Table 1.17 for the definition of the regions.)

Table 1.19 Housing mobility, percentage moving.

	1985	1989
Rented:		
below f 250	9.7	10.8
f 250–f 350	10.3	10.3
f 350–f 450	13.8	9.0
f 450–f 550	13.0	11.3
f 550–f 650		12.5
f 650–f 850	14.2	12.6
f 850 or more		15.2
average	11.6	10.9
Bought:		
below f 120,000	6.3	5.2
f 120.000–f 150,000	6.4	5.1
f 150,000–f 175,000	5.4	5.2
f 175,000–f 200,000	4.8	5.3
f 200,000–f 300,000		5.4
f 300,000 or more	4.6	5.5
average	5.9	5.2
Average	9.4	8.5

Source: NRO 1985/1989.

Table 1.20 Housing mobility by sector, percentages.

	1978		1982		1986	
	Accommodation moved to[1]					
	rented	bought	rented	bought	rented	bought
Previous accommodation:						
rented	49	40	41	32	42	38
bought	6	29	5	30	6	25
none	45	31	54	37	52	36
total	100	100	100	100	100	100

Source: WBO 1977/1978, WBO 1981, WBO 1985/1986.

Note: (1) Definition of accommodation by CBS.

Table 1.21 Net internal migration.

To:	North	South	East	West
From:				
north	–	255/720	799/1381	613/2044
south	0	–	645/0	517/0
east	0	0/224	–	0
west	0	0/96	6913/4567	–

Source: CBS, Statistisch Jaarboek.

Note: The annual average between 1980 and 1987 is shown N/. The figure for 1988 is shown /N.

It can be seen that the northern region was the biggest net exporter of population, apart from the very large flow of short-distance migration from the west to the east. In this period, net outflow from the northern region has increased, and migration *from* the south has turned into migration *to* the south.

Table 1.22 analyzes the same information in a different way, showing internal migration between different types of municipality. Comparing 1980–84 with 1985–89 shows that movements have been stable or have increased slightly; the only systematic decreases are in migrations from large

Table 1.22 Net internal migration between types of municipality.

	To:	Large cities	Small towns	Urbanized rural	Rural	Total
From:						
large cities	a	116,400	181,500	349,500	65,900	713,300
	b	132,900	174,400	347,500	44,700	699,500
small towns	a	162,300	201,500	287,100	96,300	747,200
	b	175,200	212,500	311,300	98,800	797,800
urbanized rural	a	253,300	275,900	381,500	94,100	1,104,800
	b	277,700	307,500	428,100	98,400	1,111,700
rural	a	43,200	97,600	90,300	64,000	295,100
	b	45,600	109,000	100,900	69,700	325,200
total	a	575,200	756,500	1,108,400	320,300	
	b	631,400	803,400	1,187,800	311,600	

a = the total migration in the five years 1980–84

b = the total migration in the five years 1985–89

Source: Atzema, 1991, p. 239.

The various types of municipality are described in table 1.15

Note that the category `urbanized rural municipalities' includes commuter villages

cities to other types of municipality. Reduced migration out of large cities and increased migration into them is noticeable. It is caused by changes in household structure, reduced growth of incomes, high removal costs and the wish to shorten journeys to work (Atzema 1991: 238).

Changes in social values

Many changes in norms and values have taken place in Dutch society in the past decade and have been reflected in government policy. The economic stagnation of the 1980s demanded large cuts in public expenditure, and the non-essential "soft sector" faced strong financial restrictions stemming from a "no-nonsense policy". The 1980s also saw changes in attitudes towards personal rôle patterns, marriage and cohabitation.

The environment has become a vitally important issue for the citizen and for the politician (SCP 1990: 17). It is a subject that is now taken into account in all sectors of policy-making. The connections between the environment and social, cultural, economic, planning and management factors are now being recognized (SCP 1990; WRR 1990 p. 112). An illustration of this is the question of mobility, with attempts being made to reduce the consequences of traffic on the environment and on health.

Some stubborn social problems remain. The growing numbers of those unfit for work places a tremendous strain on the social security system; while structural unemployment, particularly in low income groups and among foreign workers, combined with the problem of the poverty line, remains serious (Sec. 1.2).

1.4 The land and property markets

Land and buildings

The main features of the land and property markets are presented in outline here, as an introduction to the extensive information contained in Parts II and III. Agricultural land receives a little more attention here, because in the rest of this report it receives only scant treatment.

Most agricultural land is owned privately, the exception being much of the land in the polders, claimed from the sea. This land is owned by a public agency (Dienst der Domeinen; in the IJsselmeer polders, the agricultural land is owned by another public body, the national agency for the IJsselmeer-polders) and it is leased to private farmers. The demand for agricultural land, and hence the price, is strongly influenced by agricultural policy, especially that of the EC. For a fuller discussion of prices see Chapter 5.3.

Although most agricultural land is owned privately, and therefore bought and sold by private farmers, public administration is very active in this market as an intermediary. Perhaps the clearest example is in land-consolidation schemes. If, in a rural area, the pattern of land ownership, roads, drainage ditches, and so on. is such as to inhibit the introduction of more efficient agricultural methods, then the land can be reallocated. Farm boundaries may be redrawn, new roads built and watercourses rerouted. This can be done only if a majority of landowners agree. Then a public agency acquires the land, carries out the necessary works, and sells the new plots as agreed. The legislation under which this is currently done is the Land Consolidation Act (Landinrichtingswet, 1985). Almost the entire cultivated area of the Netherlands has been reorganized in this way at least once since 1924.

Public administration also acts as an intermediary in the rural land market in respect of afforestation and nature reserves. An agency has been set up especially for this task, the Bureau Beheer Landbouwgrond. This acts on instructions from the ministry of agriculture and fisheries, and the ministry of housing, physical planning and environment. It acquires land, manages it temporarily if that is appropriate, then disposes of it.

Finally, it should be emphasized that public policy for agricultural land is pursued almost entirely separately from policy for urban land. This policy has its own agencies and its own legislation. Apart from the Land Consolidation Act mentioned above, which is applicable only in rural areas, there are two other pieces of legislation exclusively for agricultural land: one regulates agricultural leases (Pachtwet, 1958), the other the transfer of land ownership within the agricultural sector (Wet agrarisch grondverkeer, 1981).

For the urban land market, the situation is in some ways similar. When development or redevelopment is to take place, a public agency (the municipality) fulfils a crucial intermediary function. It acquires the land which is to be (re-)developed, carries out the necessary servicing, then disposes of it to building developers. The main contrast with public administration's rôle in agricultural land is that public involvement in urban land is more intense. Almost all the land on which urban (re-)development is to take place passes through public hands (Ch. 5.2). This means (among other things) that prices for urban land are influenced much more by public involvement than are prices for rural land. Also, much of the land acquired by the municipality for (re-)development schemes is retained in municipal ownership for roads, public open space, etc. Some municipalities retain legal ownership of all the land thus acquired and dispose to the building developers only the ground leases. As a result, much urban land is in public (municipal) ownership (Ch. 8.3). Another important difference from agricultural land is the influence exercised indirectly on urban land by public administration, which is exceedingly strong, coming through the operations of physical planning, environmental policy, housing subsidies, etc.

Developed land (land plus buildings)

Most property is owned privately or by semi-public trusts. Even public services such as schools, hospitals and social housing are owned and run by semi-public trusts, not by public administration. The public administration owns only the buildings it needs for performing its public offices, i.e. town halls, administrative offices, etc. Administrative buildings needed by central government are administered by one agency, the Rijksgebouwen Dienst (Ch. 7.2).

However, although public administration has very little active involvement in the market for developed property as buyer or seller, it has a strong passive or indirect involvement through its policies for physical planning, the environment, housing, etc.

The construction industry

The construction industry is involved in the urban land market, for servicing land for new development, and in the urban property market for new buildings, restoration and maintenance. This involvement can be measured in terms of the value of the output of that industry, in Table 1.23.

But however large and indispensable that contribution is, the construction industry itself has little direct influence on the land and property markets, being more a neutral supplier of services.

The output of the construction industry is supplied by very many firms of different sorts and sizes. Table 1.24 presents information about the different sorts of firm.

The size structure of the firms in the construction industry is shown in Table 1.25

The figures in Table 1.25 show recent changes in the construction industry. The industry is very sensitive to the state of the economy (EIB

Table 1.23 Construction industry output.

Year	Output (mln Hfl. excl. VAT, 1987 prices)
1980	49,361
1982	40,837
1984	40,995
1986	43,788
1988	49,380
1989	50,630

Source: output figures taken from the national economic accounts.

Table 1.24 Construction industry output by type of firm.

Type of firm	Distribution of employment 1980	Gross turnover 1981		Distribution of employment 1983	Gross turnover 1983		Distribution of employment 1988	Gross turnover 1987	
		mln Hfl	(%)		mln Hfl	(%)		mln Hfl	(%)
Construction[1]									
building contractors	13.0	27,818	67.1	10.6	25,433	65.5	12.7	30,779	66.6
civil engineering	21.1	9,398	22.7	18.4	9,655	24.9	19.9	11,014	23.8
painters	7.5	3,177	7.7	6.6	2,899	7.5	7.6	3,269	7.1
others	8.4	1,051	2.5	7.4	814	2.1	8.5	1,120	2.4
Total		41,444	100		38,801	100		46,182	100
Installation[2]									
plumbers etc	9.0	2,236	26.0	7.8	2,025	25.0	9.5	2,524	23.6
central heating and air conditioning	22.8	2,752	32.0	19.1	3,078	38.0	19.6	3,092	28.9
insulation	16.6	688	8.0	15.4	648	8.0	20.8	755	7.1
electrical services	13.8	2,924	34.0	10.3	2,349	29.0	17.4	4,322	40.4
Total		8,600	100		8,100	100		10,693	100

Source: CBS, Statistisch Zakboek 1991. Economisch Instituut voor de Bouwnijverheid (EIB), 1991.

Notes: (1) employment in man years.

(2) employment in numbers of workers.

Table 1.25 The structure of firms in the construction industry.

Size category	1980				1988			
	No of firms	Share of workforce	Gross turnover mln Hfl	(%)[3]	No of firms	Share of workforce	Gross turnover mln Hfl	(%)[3]
Construction[1]								
less than 10	73.8	20.9	6,393	15.4	73.2	20.8	9,866	21.4
11–20	13.4	15.2			13.5	15.9		
21–50	8.8	21.2	35,051	84.6	9.1	22.7	21,426	46.4
51–100	2.5	13.2			2.8	15.6		
>100	1.5	29.5			1.4	25.0	14,890	32.2
Total	100.0	100.0	41,444	100.0	100.0	100.0	46,182	100.0
Absolute[4]	22,252	284,742			18,399	227,769		
Installation[2]								
1–9	77.4	27.9	2,494	29.0	77.2	29.0	1,942	30.5
10–49	22.6	72.1	6,106	71.0	19.7	36.3	4,430	69.5
more than 49	0.0	0.0			3.1	34.7	2,053	32.2
Total	100.0	100.0	8,600	100.0	100.0	100.0	8,425	100.0
Absolute	6,516	89,100			6,066	87,470		

Sources: CBS, Statistisch Zakboek 1991, EIB 1991.

Notes: (1) employment measured in man years.
(2) employment measured in number of workers.
(3) current prices, excluding VAT.
(4) firms with no employees are not included. In 1988 there were 15,106 such firms counted.

1991). This sensitivity is particularly important for the economy as a whole, because the construction industry is such a large part of it. The figures in Table 1.26 illustrate this.

Table 1.26 The share of the national economy devoted to construction.

Year	% share in GNP[1]	% share in value-added[2]	% share in fixed capital formation[3]
1970	16.0[4]		
1980	12.4	8.8	62.1
1982	10.5	7.3	60.3
1984	10.0	6.7	56.4
1986	10.3	7.2	51.6
1988	11.2	8.3	53.2
1989	11.0	8.3	52.2

Source: CBS, calculations by the EIB.

Notes: (1) at market prices.

(2) GNP at market prices.

(3) investment in construction works as a proportion of all fixed capital formation.

(4) source: EIB 1991.

Unit construction costs have changed in the past 20 years as illustrated in Table 1.27.

Table 1.27 Increases in unit construction costs over time.

Year	%		%		%
1970–71	11	1977–78	8.2	1984 – 85	(1.0)
1971–72	10.8	1978–79	9.1	1985 – 86	(1.0)
1972–73	9.8	1979–80	10.7	1986 – 87	2.0
1973–74	7.4	1980–81	7.5	1987 – 88	3.0
1974–75	15.2	1981–82	3.0	1988 – 89	2.9
1975–76	8.4	1982–83	(1.9)		
1976–77	7.7	1983–84	(1.0)		

Source: CBS, Maandstatistiek van de prijzen.

Average construction costs of dwellings are affected by these changes, and by changes in the size and quality of new dwellings. The figures are presented in Table 1.28. The decrease between 1980 and 1985 was caused by the very low demand for housing in that period.

27

Table 1.28 Average construction costs of dwellings.

Financial sectors	1980	1985	1987 Hfl.	1988	1989
social rented	73,290	70,130	71,780	75,250	77,460
premie	88,690	79,930	86,360	85,330	87,660
free sector	161,550	112,510	124,470	126,000	132,420

Source: CBS, Statistisch Zakboek 1991.

Notes: All dwellings for which a building permit was issued. Costs exclude VAT. The "financial sectors" of dwellings are described in Ch 7.2.

The use of space

Land use and conversion from rural to urban land During the period 1950–85 the total urban area increased by 250% (from 20.4 to 51.1km^2). This is the effect of the suburbanization process that occurred in the Netherlands after the Second World War.

The corresponding transformation of land use was for the greater part at the expense of farmland. From 1977 to 1985 the area of farmland declined by an average of nearly 5,000ha/year (annual average 0.2%). Urban development alone absorbed nearly the same area (annual average growth 0.9%). However, we must remember that a great part of the land classified as urban is, in fact, not built-up. In 1985 between 20% and 25% of urban areas consisted of "open" spaces (de Groot *et al.* 1988).

The Netherlands is a small country. In reading Table 1.29, which details land use, it should be borne in mind that a shift of 35,000ha (at first sight small) accounts for 1% of the total land area. All areas in Table 1.29 are in hundreds of square kilometres. The waters of the North Sea and the IJsselmeer which belong to the Netherlands (639km^2) are not included in it.

Table 1.30 shows how the urban area has grown and how each person on average is using more urban space.

Dwelling space per person Statistics about dwelling space per person are not collected. Moreover, this market is so segmented that an average could be misleading. For these reasons, we have to be satisfied with the figures on the average number of persons per dwelling detailed in Table 1.31.

Area of working space per person It is extremely difficult to produce reliable and meaningful figures for the use of space in industrial buildings because the requirements of firms involved are so divergent. Problems arise with regard to the description and content of the activity, the locations, the number of personnel, etc. We can indicate an average land requirement but

Table 1.29 Land use in the Netherlands.

Year	Total area abs	Built-up area abs	"Urban" recreation abs	Other urban use[2] abs	Transport abs	Total urban, transport, and other abs	%	Forest abs	Natural area abs	"Rural" recreation abs	Total forest natural area and recreation abs	%	Agriculture abs	%	Water surface width > 6 m abs	%
1950[1]	335.0					20.4	5.7	29.3	19.1	—	48.4	13.6	259.4	73.1	26.8	7.5
	348.3	17.9	0.4	—	10.2	28.5		24.3	24.6		48.9		251.2		19.1	
1955[1]	355.0					22.1	6.2	29.2	17.2	—	46.4	13.1	259.9	73.2	26.6	7.5
	351.8	17.9	0.7	0.5	10.0	29.1		25.0	22.0		47.0		252.3		20.2	
1960[1]	361.6	18.1				24.8	6.9	29.3	20.0	—	49.3	13.6	259.7	71.8	27.8	7.7
	361.3	18.1	0.8	1.0	8.1	28.0		27.0	23.0		50.0		255.9		25.0	
1965[1]	316.6	18.7				528.9	8.0	29.4	17.9	—	47.3	13.1	257.5	71.3	27.7	7.7
	360.9	18.7	1.1	2.6	6.4	28.9		28.9	19.0		47.9		257.7		26.5	
1970	366.2	21.2	1.4	3.5	6.7	32.9	9.0	29.8	19.9	—	49.7	13.6	255.2	69.7	28.5	7.8
1975	369.5	24.7	2.6	4.3	7.2	38.8	10.5	30.8	16.9	—	47.7	12.9	251.6	68.1	31.4	8.5
1977	371.8	25.8	3.6	5.3	12.7	47.4	12.7	29.6	15.6	3.0	48.2	12.9	241.3	64.7	33.8	9.1
1983	372.9	28.4	4.3	4.8	13.0	50.5	13.5	29.7	15.6	3.1	48.4	13.0	240.4	64.5	33.7	9.0
1985	373.3	28.9	4.5	4.5	13.2	51.1	13.7	30.0	15.0	3.3	48.3	12.9	239.7	64.2	34.1	9.2

Source: de Groot et al 1988, using CBS statistics from various years.

(1) original records CBS, uncorrected.

(2) e.g. future building sites, cemeteries and waste disposal sites.

The system of data-sampling in the Netherlands was revised after 1976. A warning is given against making simple comparisons over the years because of the possible changes in classifications and definitions. Nowadays the Central Bureau for Statistics is working on a digitalized system for land-use changes. Together with the change of sampling-period this is why there are no more recent figures available.

Table 1.30 The growth in urban land usage.

Year	Total urban area in m^2 (mln)	Total population per 1 January (×1,000)	Urban area in m^2/p.p.[1]
1970	3,290	12,958	254
1975	3,880	13,599	285
1980	4,860	14,091	355
1981	4,980	14,209	350
1983	5,050	14,340	352
1985	5,110	14,454	354

Source: CBS and own calculations.

Note: (1) These figures have been calculated using the simple formula: (total urban area)/(total population). Of course, some of the population will not live in an urban situation (e.g. farm-workers) : the number is unknown.

Table 1.31 Average number of persons per dwelling.

1970	3.51	1985	2.73
1975	3.18	1986	2.70
1980	2.97	1987	2.66
1981	2.93	1988	2.63
1982	2.88	1989	2.59
1983	2.83	1990	2.56
1984	2.78		

Source: CBS (from Min. VROM; 1990(c))

the problem here is the considerable range. Comparisons must only be made with caution. Only an indication can therefore be given of the space requirements for industry, for services on industrial sites and for offices.

The use of land can be presented in terms of the area coefficient (the number of square metres of site occupied by one worker), net or gross (the latter includes services, roads, public parking, etc.). Some net area coefficients are given in Table 1.32.

Another way of presenting the space requirements is in terms of the number of square metres of covered floorspace per person, net (directly used for carrying out the prime function) or gross (including corridors, toilets, services, etc.). Figures for gross area are given in Table 1.33.

An indication of the relationship between total site area, site area built upon, and floor area was discovered by Buit for industrial and wholesale companies (see source of tables): this was 100:40:52

Table 1.32 Land use by "area co-efficient'".

Type of industry	Area co-efficient net
industry (general)	500 m²
industrial estate Moerdijk	2000–2500 m²
Shell Chemicals	2800 m²
other companies on Industrial Estate Moerdijk	1000 m²
wholesale companies	200–432 m²
transport companies	250–523 m²
sewerage plants, cleansing companies	> 5000 m²

Source: T H Delft, 1981

Note: Area net coefficient equals the number of square metres of the site occupied by one worker.

Table 1.33 Covered floorspace per person.

	Local authorities with	
Sector	less than 50,000 inhabitants	more than 50,000 inhabitants
local authority services	27.4 m²	31.1 m²
banks	24.6 m²	35.1 m²
organizations	26.9 m²	34.6 m²
central and provincial government	28.0 m²	27.1 m²
trade	27.1 m²	28.2 m²
industry	26.1 m²	22.1 m²
law	23.7 m²	24.2 m²
sales agencies	20.3 m²	30.6 m²
design	22.0 m²	22.5 m²
insurance	21.3 m²	22.0 m²
administration	21.9 m²	20.0 m²
communication	19.7 m²	20.1 m²

Source: TH Delft, 1981. (The figures quoted were collected from various partial surveys. The last time national research was carried out into land use by industry was 1959.)

1.5 Trends in spatial development

Regional distribution
Table 1.34 shows population distribution between the four standard regions. The province of Flevoland sits unhappily in this division, being strongly oriented to and influenced by the western region, especially the two largest cities in Flevoland: Almere and Lelystad. Table 1.35 shows, per region, the number of employees and the industrial structure of that employment. Total employment has not grown at greatly different rates in the different regions.

31

Table 1.34 The distribution of the population between the "standard regions".

Region	1970	1975	1980	1985	1988	1990
North (Groningen; Friesland; Drenthe)	1,406	1,491	1,556	1,588	1,593	1,595
East (Overijssel; Gelderland; Flevoland[1])	2,442	2,628	2,778	2,912	2,987	3,037
West (Utrecht; North-Holland; South-Holland)	6,014	6,168	6,287	6,399	6,527	6,611
South (Zeeland; North-Brabant; Limburg)	3,095	3,312	3,468	3,554	3,607	3,649

Source: CBS, own recalculations.

All population figures ×1,000.

(1) From 1986 Flevoland has been an independent province. Before it was called "Zuidelijke IJsselmeer Polders (ZIJ and "Noord-Oostpolder (NOP)".

Table 1.35 The number of employees per region and per industry.

Industrial sector	North		East		West		South	
	1980	1988	1980	1988	1980	1988	1980	1988
0	8.6	9.5	11.4	14.3	30.9	36.9	14.7	17.9
1	3.1	4.3	1.5	2.0	1.5	2.8	1.2	0.6
2/3	106.5	96.7	211.4	202.7	403.3	356.7	328.6	331.4
4	5.2	4.8	8.1	9.3	21.8	21.2	10.5	8.7
5	45.5	32.2	86.8	68.2	178.9	150.5	99.3	86.6
6	56.2	70.6	120.0	156.2	371.8	443.1	148.5	198.2
7	21.3	25.7	36.0	45.1	171.1	200.7	47.5	61.4
8	29.2	44.5	59.6	90.7	257.9	336.5	68.7	109.6
9	120.0	171.3	217.3	304.9	610.3	821.0	252.7	348.5
total	395.4	460.4	752.1	893.2	2047.7	2369.6	971.9	1163.0
change 1980–88	100.0	116.4	100.0	118.8	100.0	115.7	100.0	119.7

Source: CBS, Statistiek Werkzame Personen.

Notes: for the definition of the standard regions, see table 1.34.

All figures for number of employees x 1,000.

The industrial sectors are: 0 agriculture and fishing; 1 mineral extraction; 2/3 manufacturing; 4 public utilities; 5 construction; 6 retail and wholesale trade, cafés and restaurants, repairs; 7 transport, storage and communication; 8 banking, insurance, commercial services; 9 other services, including the public administration.

Urban-rural and city-region distribution

Table 1.36 shows the population in different geographical types of area and how it has changed between 1970 and 1985.

Two points stand out in Table 1.36 if we look at the 1970-85 trend. First, the population has dispersed farther into the city region, especially into the urban fringe, defined by Atzema (1991) as those parts of the city region that are not in the agglomeration but that fall under the influence of the central city. Secondly, at the national scale, there has been population dispersal over longer distances, and the intermediate zone has grown in all its various parts.

Another aspect of the urban-rural relationship is daily journeys to work. Rising prosperity has led to an increase in commuting. In Table 1.37 taken from Bovy & Adel (1988), commuting is defined as a journey to work that

Table 1.36 Population in different types of geographical area

	1970 abs.(×1,000) %		1985 abs.(×1,000) %		1970-85 % change
Central zone[1]					
urban agglomeration[2]	3,846.1	29.7	3,398.3	23.5	11.6
urban fringe	1,481.5	11.4	2,097.3	14.5	41.6
city region	5,327.6	41.1	5,495.6	38.0	3.2
beyond the cities	248.3	1.9	298.9	2.1	20.4
total	5,575.9	43.0	5,794.5	40.1	3.9
Intermediate zone					
urban agglomeration[2]	997.0	7.5	1,030.7	7.1	5.5
urban fringe	1,047.5	8.1	1,467.4	10.2	40.1
city region	2,024.5	15.6	2,498.1	17.3	23.4
beyond the cities	912.2	7.0	1,138.2	7.9	24.8
total	2,936.7	22.7	3,636.3	25.2	23.8
Peripheral zone					
urban agglomeration[2]	1,344.9	10.4	1,369.6	9.5	1.8
urban fringe	1,045.1	8.1	1,272.8	8.8	21.8
city region	2,390.0	18.5	2,642.4	18.3	10.6
beyond the cities	2,052.2	15.8	2,377.2	16.5	15.8
total	4,442.2	34.3	5,019.6	34.7	13.9
The Netherlands	12,954.8	100	14,450.4	100	11.5

Source: CBS, Gemeentelijke Bevolkingstatistieken (recalculated by Atzema, 1991, p.54).

(1) The following division into zones has been used:

central: South Holland, Utrecht, the south of North Holland

intermediate: around the central zone , including the city regions of Alkmaar, Zwolle,

Deventer, Nijmegen, Tilburg, Breda

peripheral: the rest of the country.

(2) The urban agglomeration together with the urban fringe make the city region.

"Beyond the cities" is the rural zone.

crosses a municipal boundary. The number of commuters has grown more rapidly than employment. As a proportion of the number in work, this grew from 27% in 1960 to 44% in 1981 and 51% in 1985. The use of the car grew rapidly, from 8% of all journeys in 1960 to 50% in 1981 and 67% in 1985. The modal split for journeys to work in middle-sized cities for the period 1981–5 is given in Table 1.37.

Table 1.37 Journeys to work in middle-size cities, 1981–1985.

Transport mode	Within	Leaving	Entering	Entering plus leaving	All
	%	%	%	%	%
car	35	64	68	67	54
train	–	12	7	8	8
bus, tram, metro	4	10	9	10	5
bike and motor scooter	53	13	15	15	30
foot	9	1	1	1	4
total	100	100	100	100	100

Source: Bovy and den Adel, 1988.

More people are working outside the municipality where they live, and are using cars more for their journeys to work. The larger the city, the greater the ratio of inward/outward commuters. Moreover, this ratio has changed between 1981 and 1985, with inward commuting growing more than outward commuting. The larger agglomerations are becoming increasingly dependent on the smaller municipalities for their labour forces, and people who leave their own municipality daily to work elsewhere travel to municipalities higher in the urban hierarchy. Some of these statements are illustrated in Table 1.38, which gives 1985 figures.

Table 1.38 Commuting patterns in 1985.

Commuters	Four big cities[1]	Middle-sized cities[2]	Small centres[3]
living in the municipality, and working	670,800	1,455,800	2,969,600
of which:			
working in the same municipality	472,700	830,600	1,158,600
working elsewhere	198,100	625,200	1,811,300
(of these working elsewhere, those with fixed work address)	108,000	337,600	1,156,900
living elsewhere, working in the municipality	577,300	996,000	1,055,900
(of which: with fixed work address)	404,400	689,000	501,900
total working in the municipality	1,050,000	1,826,600	2,214,500
those entering the municipality and with a fixed work address/ those leaving the municipality and with a fixed work address	3.74	2.04	0.43
that same ratio in 1981	3.57	1.62	0.58

Source: CBS, Arbeidskrachtentelling 1985, own recalculations

Notes: (1) with more than 200,000 inhabitants.

(2) between 50,000 and 200,000 inhabitants.

(3) with less than 50,000 inhabitants.

CHAPTER 2
The policy environment

2.1 Physical development policy

Growth centres, and their effects

Urbanization, both of existing and of new urban areas, has received much attention in post-war planning policy. The mainstay of this policy is growth centres, a planning concept similar to Ebenezer Howard's new towns. The foundation for the policy was laid in the late 1950s, based on considerations of a rapidly increasing population, the undesirability of unchecked expansion of cities, a shortage of housing and locations, etc.

Officially, a growth centre is described as a subsidiary centre within a city region. A designated growth centre is supposed to gain at least 6,000 dwellings within ten years. More concretely, a growth centre is to contribute to the housing needs of the region by offering an attractive environment, preventing uncontrolled suburbanization, making the use of space more efficient, restraining the use of cars, offering more diversification in the housing market, and realizing simultaneously employment, infrastructure and services. The central government is responsible for growth centres. They are Alkmaar, Hoorn, Purmerend, Haarlemmermeer, Almere, Lelystad, Huizen, Houten, Nieuwegein, Duiven-Westervoort, Helmond, Spijkenisse, Hellevoetsluis, Zoetermeer and Leidschendam.

In order to implement its policy of growth centres, the national government initially assumed that the existing set of formal planning instruments would be sufficient. This turned out to be incorrect. The construction of housing began much more slowly than originally planned, the expected employment failed to materialize, and the infrastructure and services deemed necessary were slow in coming. This was because of insufficiently equipped management of the municipalities, financial obstacles, and barriers to social and economic development. The instrumental side of the policy was improved rapidly in the early 1970s. The improvements were financial (e.g. grants for the main infrastructure, subsidies and increases in the funding of the municipalities selected as growth centres), organizational (e.g. co-ordination committees), and spatial and substantive. Because of these improvements to the instruments, the growth-centre policy has become more effective since the late 1970s.

The end of the policy is in sight. Since 1985, national planning policy has clearly shifted its focus to construction and investment in or close to the large cities. The new policies are described below. There are several reasons for this: economic recession, the prospect of a stationary population (about 16 million inhabitants) and a fall in the demand for housing, the negative consequences of the overspill policy for donor cities, more residual space in donor cities than had been assumed, increasing traffic congestion and environmental problems, and less available finance.

Note that in addition to the policy of growth centres, national government has also had a policy of growth cities. There are four of these: Groningen, Zwolle, Amersfoort and Breda. Their task was to build up a new city region in a relatively empty area.

The compact city policy

In the 1980s, renewed interest in the cities became apparent. Existing cities were found to have greater unused capacity than had been assumed, partly because of the outward migration they had suffered and partly because people's wishes had changed. Furthermore, city living had become more attractive for large groups of people. Gentrification is an example of this. Also, urban renewal had improved the cities, and criticism by the donor cities of the growth-centre policy had drawn attention to social problems in the larger cities.

The compact city policy was developed in the mid-1970s and implementation started at the beginning of the 1980s. This term is generally used to refer to a utilization of possibilities in an existing urban area and the maintenance of its environmental quality (Jansen et al. 1985).

Features of this policy include maintaining and improving the functioning of the cities as a good environment for living and working; utilizing existing capacity for urban functions, and mixing and combining functions so that they support each other; utilizing existing capital investment; stimulating the use of bicycles and public transport; maintaining a viable size to support urban functions; using space as intensively as possible; and restriction of urban development outside the cities. The stress laid on both quantity and quality of urban development (for example, not filling up all open spaces) and on the fact that dwellings alone are insufficient without employment, services, recreation, transport, etc. is also important.

Urban renewal policy

In the 15 years up to 1985, when urban renewal legislation came into force, central government pursued an explicit urban renewal policy using existing laws. Until 1985, the legal basis was provided by existing legislation.

Government policy was to give subsidies to urban renewal projects, and there were more than 40 separate grants. Applications were tested centrally and in this way national government had a decisive influence on the total policy.

Under the Town and Village Renewal Act 1985 (Wet op de Stads- en Dorpsvernieuwing), an important part of that decision-making was decentralized (see Ch. 2.6). The money is no longer allocated to specific projects, but annually in blocks to municipalities and provinces from an urban renewal fund according to objective rules. These authorities are then free to decide how to use the income as long as it is allocated to renewal. For this purpose, municipalities are divided into four groups: (a) the "four large cities"; (b) other municipalities included in the national urban renewal programme (Meerjarenplan Stadsvernieuwing or MPS); (c) other municipalities receiving grants directly from central government; and (d) municipalities receiving the grants via the provinces. Groups (a) (b) and (c) contain 89 municipalities and receive 81% of the grants.

According to the MPS for 1990-4, the main objective is "adapting and intensifying the use of the environment in the areas built before 1971 for living, working and producing so as to meet the present wishes and standards in such a way that the arrears are made good by a radical improvement, all for the benefit of those who live, work, or manage a firm in those deprived areas" (Min. VROM 1989a). Clearly this entails more than routine management and maintenance.

Over the years, both national and municipal governments have shifted their emphasis in urban renewal. A policy of "renewing for the neighbourhood" has been replaced by a wider policy of "renewing for the city". This can be seen in the approach to firms in renewal areas. Instead of helping each firm separately, the economic structure of the city as a whole is now considered.

Municipalities are now taking a wider view. It is not only the housing stock dating from before 1939 that is being tackled, but also that built more recently, and especially that in the private rented sector. It is not only quantity that counts, but also the quality of the residential environment. Measures for public open space, traffic safety, environmental protection (e.g. against contaminated ground) are now crucial. Better connections are also being made with policy in other fields, such as education, welfare and environment.

In summary, central government creates the conditions (money and powers), the municipalities make and implement policy, and the provinces support and supervise the process.

The policy for urban revitalization

The Vinex report (Min. VROM 1990b) described urban revitalization in these terms: "the Cabinet lays emphasis on achieving good living and production surroundings, on utilizing the available capacity in urban areas for living, working, recreation, and services, and on mixing these functions, all with the aim of maintaining and improving the physical conditions necessary for the city to function well".

In practice this means trying to realize ambitious urban redevelopment schemes by means of public-private partnerships (PPP). In this way, the cities hope to present themselves as irresistibly attractive to firms, institutions and households.

Inter-urban competition is stirred up by central government. In its Fourth National Report on Physical Planning (1989) the term urban intersection was introduced, and the following nodes were designated. Amsterdam, Rotterdam and Den Haag were deemed to be of international significance; Utrecht, Arnhem/Nijmegen, Eindhoven and Groningen of national significance; Enschede/Hengelo and Maastricht/Heerlen of regional significance within Europe; and Leeuwarden, Zwolle, Breda and Tilburg of regional signifi-cance.

The aim is to select thriving cities that will then receive support from central government to maintain and strengthen their positions. Such cities get priority in central government investment. Private investors are sought, to complement government support. In practice, it is hoped that they will provide most of the necessary investment.

Another aspect of urban revitalization is the "exemplary redevelopment schemes" in which various types of arrangements between public and private partners are being tried out. The first of these are under way in Maastricht, Amsterdam, Rotterdam, Den Haag and Groningen. Public–private partner-ships can be made outside the "urban intersections", but only the designated locations will get extra financial support from central government.

Geographical concentration of employment

Mobility is a part of physical planning policy currently receiving much attention. As early as 1983, in the Structuuurschets Landelijke en Stedelijke Gebieden (urban and rural areas), locational criteria were formulated in connection with this issue. A solution is now more urgent because of deterioration of the environment and of accessibility.

The present aim, as expressed in the Vinex report (Min. VROM 1990b), is to reduce the need to travel and the distance travelled, especially by cars and lorries, by influencing the choice of location and by stimulating the use of bicycles and public transport.

One of the ways in which this is being pursued is a locational policy for firms and services. The basic principle is the right firm in the right place. This is worked out in the "ABC formula", as shown in Table 2.1.

Using the criteria laid down in Table 2.1, a connection is made between the mobility profile of firms and services and the accessibility and nearness to concentrations of population. In particular, firms of type C will not be expected to locate in cities and may locate far away from them and near motorways. On the other hand, firms of type A will be expected to locate next to a node of public transport.

Table 2.1 Locational policy for firms and services.

	Type A firms	Type B firms	Type C firms
Space requirement per worker	less than 40m^2 per employee	40 –100m^2 per employee	more than 100m^2 per employee
Requirements of visitors	daily stream of visitors, reception function, etc. (Less than 400m^2 floor space per visitor)	regular contact with clients and customers who visit the firm. (100–300m^2 floorspace per visitor)	occasional contacts with clients and customers who visit the firm. (More than 300m^2 floorspace per visitor)
Dependence of business activities on use of car	less than 20% of personnel have to use the car	20–30% of the personnel have to use the car	More than 30% of the personnel have to use the car
Importance of motorway connection for goods transport	hardly important	possibly important	important

Source: Min. VROM, 1990(b), p. 18.

2.2 Regional social–economic policy

Regional socio-economic policy enables central government to influence the distribution of population and economic activities between the regions of the country. Such a policy can have important effects on land and property markets.

In the past, such a policy was actively pursued. Objectives included prevention of congestion in areas growing most strongly (i.e. the Randstad in the west) and the reduction of regional inequalities, especially in respect of higher unemployment rates in the north and south. The measures used were familiar: giving grants to private industrial investment in development areas and funding investment and social programmes there. Powers to prevent or discourage industrial building in certain congested areas were made available (*Selectieve investeringsregeling*). A permit was sometimes necessary; sometimes there was an obligation to announce expansion plans or even the imposition of a levy, although the latter was rare. In the 1980s, this policy was steadily abandoned. Emphasis came to be placed instead on economic growth, which implied that it take place in regions best suited to it. In practice, this often means the Randstad area, especially now that physical planning policy is encouraging growth in existing large towns and cities, under the compact city policy explained in Chapter 2.1.

Officially, the regions are now encouraged to pursue their own economic development policies, with limited financial assistance from central government in the form of grants (IPR) towards industrial investment in a few areas with structurally weak economies in the east, north, and south. Its highest value is 35% of the cost of a project, but it is not given for investment in buildings for rent. There is no restriction (apart from normal planning and environmental controls) on economic development in the rest of the country.

2.3 Policy for the building industry

An important policy instrument for the building industry is the construction programme (*bouwprogramma*), which the minister for housing, physical planning and environment presents to parliament every year. This obligation is laid down in legislation passed after the Second World War for the restoration of the whole society (Wederopbouwwet). The initial objective of the construction programme was to match scarce production capacity and finance to priorities fixed by the government. This was necessary because of the huge shortage of building works caused by stagnation of production and destruction during the Second World War.

That shortage was gradually removed, and the construction programme now has a different objective. It is no longer so important to allocate scarce capacity; the aim now is to ensure continuity for the building industry. This industry is very sensitive to macroeconomic fluctuations, so it is desirable that changes are foreseen well in advance. The present construction programme is an estimate of monetary demand for the output of the construction industry in the middle term.

2.4 Housing

Building to meet housing needs

Section 22, paragraph 2, of the Constitution reads: "It is the responsibility of government to see that sufficient housing is available". In the post-war years the main emphasis was on building as many new dwellings as possible. As a result, housing in the Netherlands greatly improved. As affluence came, people became more demanding about the space they needed and the quality of their homes. Nowadays there is not an absolute housing shortage. The apparent shortage is more a statistical one, a feature of certain sections in local housing markets.

The new trends mean that quite a substantial number of new dwellings will still have to be built, increasing the stock of dwellings to 7.3 million in 2025 (Min. VROM 1989c). At the same time, more attention will have to be paid to the existing housing stock, which will increasingly gain more significance than the building of new dwellings. The total number of new dwellings built is expected to drop from 94,500 in 1988 to 80,000 in 2000. In that same period, the percentage of private sector housing in new building should rise from 20% to 50%.

The present policy is to ensure that sound and reasonably priced accommodation is available for the lower paid (see below and Ch. 3); to promote a good living environment; to encourage home ownership; and to encourage experiments, innovation and transfer of knowledge in, for example, durable construction.

Special attention is given to households (more than one person) with an income below 30,000Hfl per annum, or single-person households earning below 22,000Hfl per annum. These are people who would be unable to find decent living accommodation without government support, and so subsidies are concentrated upon them. A big obstacle to this policy is the poor allocation of households over the stock. Many households with above average incomes live in inexpensive dwellings. The last Housing Needs Survey (WBO) in 1985 revealed that this was the case for almost one-third of the cheaper rented dwellings (below 450Hfl per month), accommodating about 750,000 people. Conversely, more than 50% of all rented accommodation above 600Hfl per month was rented by households with lower incomes (Min. VROM 1989c). A rectification of this misallocation and a strict policy on new allocations are important, but because of the variety of local conditions this is a task for local authorities, although central government gives financial support (see Ch. 7. 3).

Improvement and adaption of dwellings

More attention is to be paid to the existing housing stock, which is ageing and deteriorating. In the past, the importance of management, maintenance and replacement was not sufficiently acknowledged. A concerted effort has to be made to catch up. The condition of the housing stock is discussed in Chapter 8.1.

The programme for subsidized housing improvement is designed to eliminate by 2000 the backlog in the renovation of the pre-war housing stock and half of the backlog in the dwellings built between 1945 and 1968 (Min. VROM 1989c). If possible, links will be made with urban renewal and environmental policy, for instance by subsidizing energy-conservation measures in houses. This is, or will become, one of the main objectives of national housing policy.

Encouragement of owner-occupation
The percentage of owner-occupied dwellings in the Netherlands has traditionally been lower than in most other West European countries. In 1985 they accounted for 43% of the total stock. The government wants to encourage the growth of home ownership, to reach 50–55% by 2000. An increase in owner-occupation will largely occur through new construction, especially of more expensive dwellings and those costing between 130,000Hfl and 200,000Hfl.

Central government has three important instruments designed to promote owner-occupation: means-tested purchase grants to offer target groups the opportunity to own their own homes, be they newly built or existing; raising of rents for social housing faster than inflation; and the sale of subsidized rented accommodation to occupiers. Tax rules are also very important for growth in owner-occupation: interest on mortgages will remain tax deductible (see Ch. 7.3).

2.5 Policy for the environment

The national environmental policy of the Netherlands is expressed in the National Plan for the Environment or NMP (Min. VROM 1989b) and in the NMP-plus (the supplement, Min. VROM 1990a). It is based on the social and political recognition of environmental issues and the reports of the Brundt-land Commission (*Our common future*, 1987) and the RIVM (*Zorgen voor morgen – Care for the morrow –* 1988, an analysis of environmental problems). The main aim is "the maintenance of an environment which can cope with continuous development" (NMP, Min VROM 1989b: 15). Some of the generally accepted starting points are the "stand-still" principle, attacking the source of environmental problems; the polluter pays principle; the use of the best available technology; and a propaganda war to promote environmentally friendly behaviour.

The enormous number of regulations that can be applied can be split up into those that deal with sources, effects, finance and other areas. Source measures include norms for emission, efficiency of production of energy and raw materials, ABC locations (Ch. 2.1), and home insulation and energy conservation. Effect measures are applied when controls cannot (yet) be exercised upon sources. Examples include traffic noise barriers, the possible use of offices for this purpose along roads and railways, and zoning of industry. Finance is used in many ways as a stimulus: as subsidies, grants, taxes to control consumption (e.g. tax on fuel to reduce car use), taxes on final use to recover costs (e.g. on the costs of water purification). Other areas include penalties, information and research.

These measures are operated at national level through the legal system and by gentlemen's agreements (covenants) made between government and industry. The provinces and the local authority also have an important function.

2.6 General

Reduction of public expenditure
From the beginning of the 1980s it became a political priority to reduce the public-financing deficit. A no-nonsense policy was adopted to make stringent cuts in public expenditure. The aim was that the same effects should be achieved if possible, but for less money. In addition, various tasks were (and are still being) decentralized and privatized.

The figures in Table 2.2 illustrate the situation. The policy goal is expressed in terms of reducing the government deficit as a proportion of the GNP. Clearly, success is partial and central government expenditure has not declined absolutely.

Decentralization of certain public administration functions
In the 1970s, central government took on more and more responsibilities. This was based on the idea that a desired society could be created by planning and regulation. However it became apparent that central government could not do this. Now the government sees a more modest rôle for itself, simplifying regulations, reducing the need for monitoring and avoiding bureaucratic red tape. This is reflected in decentralization and deregulation. The idea is that the carrying of responsibility, the powers, and the running of risks should be combined in one agency wherever possible (Min. VROM 1989c). This is why the 1990s will see a reallocation of responsibilities. In housing, for example, the main emphasis will be on individual responsibility.

Table 2.2 Government debt and GNP.

	1975	1980	1985	1987	1988	1989	1990[1]
Central government balance of public finance (mln Hfl)	(6,417)	(15,282)	(23,071)	(15,463)	(22,180)	(22,876)	(23,233)
% of national income	3.2	5.0	6.1	4.0	5.6	5.4	5.2
taxes received per capita (Hfl)	4,349	6,601	7,035	7,971	8,300	8,333	9,608
national debt per capita (Hfl) (31 Dec.)							
central government	2,906	6,545	15,109	15,888	17,203	18,749	20,679
other authorities[2]	3,684	4,625	5,048	5,769	5,722	5,593	5,519
total	6,590	11,170	20,157	21,657	22,925	24,342	26,198

Source: CBS, Statistisch Zakboek, 1991.

(1) provisional and/or estimated figures.

(2) excluding central government loans for housing policy.

This applies to the lower tiers of government (provinces and municipalities), to the housing associations and other interest groups, and to the housing consumer.

This change can also be seen in nearly every other aspect of public administration. One of the best examples is the decentralization and deregulation within the new (1985) legislation for urban renewal discussed above (Ch. 2.1). The grants go straight into an urban renewal fund and the local and provincial authorities decide upon its allocation (Min. VROM 1989a, Schuiling 1987). No more is regulated than is strictly necessary.

Note also that decentralization is often carried out by delegating tasks but not the necessary finance. It is a hidden way of economizing.

Privatization

Towards the end of the 1980s, great stress was placed on privatizing certain public tasks, partly as a way of reducing public expenditure. Certain tasks were fully privatized, others were to become self-financing, with the user paying the full costs. For urban development and redevelopment, links with the private sector are increasingly sought. Examples include the encourage-ment of owner-occupation, the sale of subsidized rented dwellings (but only under restrictive conditions), and public–private partnerships in urban revitalization and in certain infrastructure works such as tunnels.

PART II
The urban land market

CHAPTER 3

The framework within which the urban land market functions

3.1 The legal environment

Legislation

The statutory system of physical planning This section describes how the public authorities try to influence the nature, form and location of physical development by applying the Physical Planning Act (Wet op de Ruimtelijke Ordening), first passed in 1962 and substantially modified in 1985. This legislation is intended to be applied in combination with the Housing Act (Woningwet) first passed in 1901: the Physical Planning Act on its own gives few powers for implementing physical planning policy. The description given in this section is brief, because detailed descriptions in English have been given elsewhere (eg Bruil et al. 1987, Dekker 1991, Spit 1987, Brussaard 1987). These sources have been used, sometimes verbatim, in the following.

A good and very practical start to a description of the way in which physical planning seeks to guide urban development is to take the viewpoint of the person who wants to change it. How is his freedom to do this influenced by the planning system? The crucial point is that certain building works may not take place unless first authorized by a building permit (*bouwvergunning*). This is regulated by the Housing Act, not the Physical Planning Act. Under a change introduced in 1992, building works are divided into (a) free, i.e. exempted; (b) those that must be reported; (c) those for which permission is required. The would-be developer has to apply to the municipal executive for a building permit. This application is tested against four sorts of standards:–

○ Does it comply with building regulations (the technical regulations relating to structural safety, etc)?

○ Does it require a permit under the Protection of Cultural Monuments Act (Monumentenwet) or under a provincial or municipal ordinance?

○ Does it conflict with the local land-use plan (*bestemmingsplan*)?
○ Since the Town and Village Renewal Act came into force (1985), another test is whether the building permit would contravene the regulations of a "living conditions ordinance" issued on the basis of the Act (see below).

If the application fails on one or more of these grounds, then it *must* be refused. And if it meets all four, then it *must* be granted. The municipal executive must give its decision about the application within two months, a period that can be extended only once and by at most two months.

The building regulations (the first test) are largely concerned with technical and hygienic matters but must also contain a clause that the appearance or visual aspects of the proposed development be tested. This is done by an independent body of experts (*welstandscommissie*), which advises the municipal executive.

The third test requires the application to be tested against the land-use policy of the municipal authority. It is therefore akin to development control, but there is no separate planning application as such. There are two exceedingly important aspects of this test against land-use policy:

○ If there is no valid land-use plan in force, or if the municipal council has not taken a formal decision to start preparing one, the planning test cannot take place. However, most municipal regulations make it unlawful to use premises for other than their existing use. Even in areas where there is no land-use plan in force,this means that there can be no change of use without a permit.
○ If the application conforms to the land-use plan, then the application *must* be granted; and if the application does not conform to the land-use plan, then it *must* be refused. (This is expressed by saying that the land-use plan is legally binding.)

If there is a valid land-use plan covering the land on which an application for development is made, then an additional type of permit, a construction permit (*aanlegvergunning*) might be required. The plan may stipulate that within an area indicated in the plan it is prohibited to execute certain works, not being building works, without a construction permit. Such a permit may be required only in so far as this is necessary to prevent a piece of land from becoming less suitable for the realization of the land use designated in the plan, or in order to maintain or protect an already realized land use. The construction works thus regulated are those affecting ground levels, drainage, planting, etc.

The would-be developer must therefore know three things. Is there a valid land-use plan covering the land on which he wants to build? If so, what is its content and who determines this? And how flexible is this legally binding document?

As regards the first question, a municipality is obliged to make a land-use plan for that part of its territory falling outside the built-up area (such areas are usually covered by several such plans). Within the built-up area, a land-use plan is not obligatory, but most municipalities have made such plans for part or all of this area.

As regards the second question, it is the municipality that makes the land-use plan, but its freedom to determine the content can be constrained in the following ways:

○ the plan must be approved by the provincial executive;

○ the municipality may choose to make a *structuurplan*, a land-use plan for part or all of its territory. This plan is not legally binding and its main function is to set a policy framework for the land-use plan. It is not a legal requirement that the land-use plan be consistent with the *structuurplan*, but if it is not, then the chances of the land-use plan being approved by the province and endorsed by the municipal council are reduced;

○ the province may choose (and all do) to make a *streekplan*, a land-use plan for part or all of its territory. If such a plan is in force, the province, when deciding whether to approve the land-use plan, will take into account its conformity with the *streekplan*. The provincial executive has the power to impose a directive (*aanschrijving*) upon a municipality in order to make the land-use plan conform to the provincial planning policy;

○ the national government may take physical planning key decisions about some aspect of the physical development of the country. A land-use plan is seen by a regional agency of national government before being approved by the province; if this agency considers that the land-use plan is not in conformity with national planning policy, it will advise the province accordingly. It can happen, nevertheless, that the municipality wants to do something inconsistent with national policy (e.g. when that policy requires that the municipality amend its land-use plan, the municipality takes no steps to do that). National government can, with a directive (*aanschrijving*), impose a certain content on the land-use plan. But this is a power that it uses with great reluctance and very rarely.

The third question is about the flexibility of the land-use plan. The statement that a land-use plan is legally binding can give the impression that it is inflexible. However, flexibility can be built into each particular land-use plan. Also, the system itself has a certain flexibility, irrespective of the particular land-use plan (Verhagen, 1989).

Flexibility can be built into a particular land-use plan in five ways:

○ the plan can be drawn up in global terms, with the filling-in of details delegated to the municipal executive;

○ since 1985, the plan can be drawn up in global terms, without the obligation that it ever be worked out in detail;

○ certain aspects of the plan can be amended, the aspects and the scope for amendments being specified in the plan;
○ the municipal executive can be authorized to grant (small) exemptions from certain (specified) aspects of the plan;
○ the municipal executive can be authorized to impose conditions additional to those specified in the plan, although these additional conditions may only be slight. (This power increases the flexibility with which the municipality can implement the plan, but not the flexibility with which the citizen may deviate from it.)

The whole system of statutory town and country planning has, in addition, certain components that may be used to give flexibility to *all* land-use plans. These are:

○ the land-use plan can be revised or withdrawn. The whole plan must be revised after 10 years, but partial revision is also possible and can be used as follows. An application is submitted that is inconsistent with the land-use plan: nevertheless, the municipality wants to be able to grant it, so it decides on a (partial) revision. The procedure, however, lasts several years;
○ the municipal executive can grant exemption from the conditions in the land-use plan for a limited number of years. This can be used to allow temporary uses;
○ the municipal executive can also grant exemption permanently, but only if the municipality is in the process of making a new land-use plan to replace the existing one and if the proposed development would be consistent with the plan being prepared. This clause in the legislation (Physical Planning Act, Article 29, previously Article 19) has been widely abused by municipalities that wanted to approve development that the valid land-use plan would have forbidden. In the 1985 revision of the Act, the procedures for use of this article were tightened up, and a legally safer way of achieving extra flexibility was created, avoiding the need to abuse it;
○ if the applicant has his application for a building permit refused by the municipal executive, he may submit an objection to the municipal council against this decision. Thereafter, he has the option of appealing under the Administrative Justice Act (AROB). Appeals are rarely used as a way of getting permission for development that contravenes the land-use plan.

The Town and Village Renewal Act The Town and Village Renewal Act (Wet op de Stads-en Dorpsvernieuwing, 1985) can affect the functioning of the urban land market because it creates three instruments for regulating development in certain areas:
○ the urban renewal plan (*stadsvernieuwingsplan*). This has the same status

as a land-use plan (*bestemmingsplan*) and is subject to the same procedures. It has much closer links with implementation, however, enabling, for example, compulsory purchase to be used for modernizing and replacing buildings. Moreover, the plan has to be accompanied by an implementation plan in which costs, returns, financing, sequence, completion date, rehousing of residents and public involvement are considered. The municipality must draw up rules for public participation.

○ the living-conditions ordinance (*leefmilieuverordening*). These have a duration of up to 10 years and can be used to combat deterioration in the living and working conditions in a given area by forbidding certain forms of land use. The ordinance is an extra ground against which the application for a building permit can be tested, alongside those described above. The applicant can be required to deposit a sum of money in the bank, to increase the certainty that the permit will be taken up. The municipality can oblige the private land owners to permit temporary facilities (eg planting, parking, playgrounds) on their land. The living-conditions ordinance cannot be used for compulsory purchase. They acquire legal status more quickly than does a land-use plan because the procedure is less complicated.

○ the right of first refusal (*voorkeursrecht*). With this instrument, the municipality has first refusal of land offered for sale in areas specifically designated for that purpose in structure and land-use plan areas. The seller informs the municipality, which has two months to decide whether to buy. Subsequently, the price is determined by negotiation or in the courts.

Environmental protection Certain legislation passed for environmental protection can also have significant effects on the urban land market.

The Public Nuisance Act (Hinderwet, 1952) requires that activities that can cause danger, damage, or nuisance must obtain a permit to operate, to expand, or to change the operating processes. Most industrial production falls under this. In particular, air pollution by smaller firms is regulated by this legislation. It also covers nuisance from activities such as cafés, dance halls, intensive farming, motorbike scrambling and shooting ranges. The municipality is the implementing authority, and it can pass local ordinances regarding nuisance. The Public Nuisance Act is often applied in combination with land-use plans (*bestemmingsplannen*) for industrial estates, as described later.

The Noise Nuisance Act (Wet geluidhinder 1979) must be seen as a complement to the Public Nuisance Act. The following parts are relevant to the urban land market:

○ noise (buffer) zones must be drawn around industrial estates and military exercise grounds, and along roads and railways. Within these zones, new noise-sensitive land uses (in particular housing) may be developed only under certain conditions. Beyond the edge of the zone, the noise emanating from the noise generator must not be above a certain value. These conditions within and beyond the zone are described in terms of noise-nuisance norms specified in the Act. The requirements must be satisfied by the content of the land-use plan and before a building permit may be granted. Also, the existence of a noise zone can be used to refuse further development within the area generating the noise. In those ways, implementation of noise zones is a matter for the municipality.

○ large industrial plants must apply for a noise permit from the province.

○ The Protection of the Ground Act (Wet Bodembescherming 1986) aims to protect soil and ground water against pollution, and is implemented by the province. Moreover, it can overrule the Public Nuisance Act. The Protection of the Ground Act can affect urban development because a firm establishing in an area declared to be a water extraction zone might have to take extra precautions to avoid dangerous substances entering ground water.

○ The Interim Act on Soil Decontamination (Interim Wet Bodemsanering) regulates the cleansing of contaminated soil. As such, it should help the urban land market to function better, because soil so badly contaminated that there is no demand for it can be brought back into supply using subsidies under the Act (for details see Ch. 3.3). Part of the subsidies can be recouped from the polluter. If the land with the contaminated soil is bought by a person other than the polluter, the new owner is not responsible for the decontamination costs, unless he has signed a sales contract agreeing explicitly to take over all such environmental claims. In many cases it is advisable that the potential buyer of land has soil investigations made before signing the contract (see *Tijdschrift Milieu Recht*, 1990).

Since 1987, environmental impact assessment (EIA) has been regulated within the Environmental Protection (General Provisions) Act (Wet Algemene Bepalingen Milieuhygiëne, 1980). This is the way in which EC Directive EEC/85/337 has been incorporated into Dutch law, although the regulations go further than the directive requires. The regulations require that procedures that give environmental aspects an appropriate place in planning and decision-making processes are followed. The procedures do not replace existing decision making, but do influence it. The administrative order lists the activities and plans for which an EIA is required, mostly by specifying a size beyond which the EIA is necessary. Examples include motorways, railways, harbours over 100ha, airports, military training areas over 100ha, urban expansion schemes with more than 2,000 houses, quarrying and excavation over 100ha, industrial areas over 100ha, and power stations producing over 300MW. If an EIA is required, the EIA procedure usually runs parallel to the prescribed procedure for making a formal decision about the activity, such as issuing the building permit.

Protection of undeveloped areas from urban development

The way most usually followed for protecting rural areas is the land-use plan (*bestemmingsplan*). Such a plan is obligatory for all space outside built-up areas, but not inside them. As this type of plan is binding, an area designated in the plan as remaining in rural use usually does so, unless the land-use plan is formally changed. A land-use plan of this type is sometimes used to implement a buffer zone, and is quite effective in doing this.

Two other items of legislation can be used to protect rural areas. The Forestry Act (Boswet, 1961) applies outside built-up areas, but not to trees in yards and gardens and alongside roads, nor to fruit trees and some other categories. It requires advance notice to be given of intention to fell trees; to do so without permission is an offence. Even if permission is granted, an obligation to replant in the same place may be imposed.

The Nature Conservancy Act (Natuurbeschermingwet 1967) gives the minister of agriculture and fisheries and the minister of housing, physical planning and environment joint power to designate an area as a protected natural monument. The legal effect of this is to make it an offence to commit or condone acts that damage the natural values or scientific significance of that area, without prior permission of the minister. When an interested party suffers damages as a result of designation or of permission being refused, the minister must pay fair compensation. All the various ministerial decrees are subject to appeals to the Crown.

The areas designated as protected natural monuments remain in private ownership. The Nature Conservancy Act also makes it possible for central government to acquire areas with a high natural value (state natural monuments: *Staatnatuurmonumenten*). The central government, as land owner, can under private law protect that land from urban development.

The process of making and adopting local plans

First, the formal procedures for land-use plans (*bestemmingsplannen*) will be described. Secondly, and in so far as they are different, the procedures under the Town and Village Renewal Act are treated. The procedures for the other types of plan-instruments – structure plan, regional plan, physical planning key decision – will not be considered here, because their effect on the urban land market is indirect and, legally at least, only through the land-use plan.

The land-use plan Initiatives for making land-use plans almost always comes from municipalities themselves, although both provinces and central government are empowered to issue a directive to a municipality that it must make a new land-use plan or modify an existing one. The decision to make a land-use plan carries the obligation to carry out research into the existing situation and into the possible and desirable development of a municipality. The research must also consider a plan's feasibility. The importance of the conclusions about financial feasibility will become clear in Ch. 3.2.

A municipal council can choose to issue a preliminary decree that a land-use plan is being prepared for the area indicated in the decree. The aim is to be able to defer decisions on applications for building permits or to require that construction permits be applied for, so as to be able to prevent developments that might jeopardize the forthcoming land-use plan. If, however, the draft land-use plan has not been put on display within one year of the decree coming into force, the decree lapses.

A draft land-use plan must be put on display (open for public inspection) for one month. During this time, anyone may lodge objections with the municipal council. After the public display period has ended, the municipal council has two months within which formally to adopt the plan; if there have been objections it has four months. The formally adopted plan is then put on display for a further month. During this time, objections may be lodged with the provincial executive by those who objected to the draft plan or by those who object to amendments made to the draft plan before it was adopted.

After the second public display period, the provincial executive has three months within which formally to approve the plan. If there have been objections, this period is extended to six months. The provincial executive may approve the plan in its entirety or partly. When it approves parts of the plan against which no objections have been lodged with the provincial executive, those parts thereby become irrevocable.

Within one month of reaching its decision, the provincial executive must make it known. The plan as approved is then put on public display for a period of one month. During this time, appeals may be lodged with the Crown:

○ by those who submitted objections to both the municipal council and the provincial executive;
○ by those who objected to amendments made to the draft plan before it was adopted;
○ by anyone who objects to the withholding of approval by the provincial executive;
○ by the inspector of physical planning (an official of the central government).

Appeals are not possible against those parts of the plan which have become irrevocable (see above).

If an appeal has been lodged with the Crown, the council of state (department of administrative disputes) has to give an advice to the Crown within twelve months. If the Department is asked for further advice after this, it has to give it within six months. Note that the appeal is not decided by the minister for physical planning.

The Crown then has to decide within six months, a period that can be extended for one period of three months. The minister publishes the decision "without delay".

The land-use plan has then acquired legal status. It is stipulated that it be revised after ten years, but it remains in force after that date unless formally withdrawn, modified, or superseded.

It will be seen that no maximum duration of time is specified for making the plan, i.e. for the period between initiating the plan-making and publishing the draft plan, unless a preliminary decree has been issued. However, after the draft plan has been put on public display, procedures come into force that can last as long as 9 months if there are no objections or appeals, with up to another 32 months if there are.

After the land-use plan has acquired legal status, it is not immutable. Flexibility can be built into a particular land-use plan as described above. The Physical Planning Act also allows a particular land-use plan to be changed. The possibilities are full revision, partial revision, and revision preceded by anticipation. For each of these possibilities, the procedures to be followed are legally prescribed. The details of these procedures may be found in English in Bruil et al. (1987), and Brussaard (1987).

The Town and Village Renewal Act The urban renewal plan made under the Town and Village Renewal Act has to follow the same procedures as are specified above for the land-use plan. Where living-conditions ordinances are made under this legislation, the procedures are simpler, so take less time. One reason why such regulations are made is that they acquire legal status more quickly than does a land-use plan, and so can be used to intervene quickly in an urban area that might otherwise deteriorate rapidly.

Public participation The possibility of members of the public influencing the content of land-use plans by lodging objections (with the municipal council and the provincial executive) and appeals (with the Crown) have been described above. In addition, the public has the right to be consulted while the draft plan is being prepared.

The Physical Planning Act, as amended in 1985, introduced the requirement that a municipality must establish a public inquiry ordinance (Article 6a). Such an ordinance must contain rules about how citizens, and those natural and legal persons with a material interest and living in the municipality, will be involved in the preparation and revision of physical plans. How the plan is to be made public, the conduct of consultations, the manner in which the findings are reported, the way in which complaints about the conduct of the consultations are handled, all have to be covered by this ordinance. It is intended that this ordinance, which is specific to physical planning, should be superseded by a general ordinance regulating public participation for many aspects of a municipality's work.

Private law relating to land and land transactions The most important parts of private law relating to land and land transactions have already been described above. Three points need emphasizing here because of their relevance for the urban land market.

When legal ownership of a piece of land is transferred, the existing owner may include in the contract both restrictive covenants and positive obligations, which apply to the new owner.

The existing owner may also include in the sales contract conditions that are binding not only on the new owner, but on all subsequent owners. Such conditions are limited to restrictive covenants and to permitting another to do things that affect the new owner's ownership rights.

The legal owner of a piece of land can grant to another a ground lease on that land. This is described in the (new) civil code, Book 5, title 7. Article 85 says: "A ground lease is a real (or material) right that gives the lessee the right to hold and to use someone else's landed property". In the contract establishing this, the lessor can lay upon the lessee the obligation to pay a ground rent.

Neither the civil code nor other legislation imposes many restrictions on the conditions that may be included in the ground lease: that is largely a matter for the two contracting parties, lessor and lessee. The contract usually regulates such conditions as the duration of the ground lease, the level of that lease and revisions to it, how the contract can be terminated prematurely, what happens when the contract has expired (e.g. does the lessee receive compensation for the remaining value of what he has invested in the site), what the site may be used for, obligations to maintain any building constructed on the site, modifications to the level of the ground lease should the land use or the building intensity be changed, etc.

Instruments for implementing plans under public law
The most important instrument for implementing plans under public law is the building permit. The relationship between this permit and the land-use plan has already been described.

The construction of buildings is allowed only in conjunction with the issue of a building permit by the municipal executive. The grounds for refusing such a permit are as described above. No other grounds are possible. This is regulated in the Housing Act (Woningwet), those parts of which relating to the building permit have recently been substantially revised, coming into force in 1992. This legislation also defines "building activity" and specifies that municipalities must apply building regulations. Drawing up those regulations was formerly a task for the municipality, and most municipalities adopted the Model Building Regulations (Model Bouwverordening) drawn up by the Dutch Association of Municipalities (Vereniging van Nederlandse Gemeenten). Under the revisions of 1992, all municipalities have to apply the same set of building regulations, determined nationally. In these latter, a building is defined, the technical and hygienic criteria against which the building permit must be tested, and the requirements to which the application for the building permit must conform.

If an application for a building permit is refused, the applicant can lodge an objection with the municipal council. If this is refused, he may appeal further, but only on procedural grounds under the Administrative Justice Act (AROB, Administratieve Rechtspraak Overheidsbeslissingen). Third parties materially affected by an application may object to the municipal executive. If that does not achieve the desired effect, they may appeal, but only on procedural grounds, under the Administrative Justice Act.

If building has taken place without a building permit or in violation of it, the usual step is to induce the offender to apply for the necessary permit. If this is then refused, the municipality may use executive enforcement (*bestuurlijke dwang*), demolishing the structure at the expense of the offender, thus re-establishing the previous situation. The municipality may also start criminal proceedings, but does so rarely. As the situation causing offence remains unchanged both during and after such proceedings, no improvement results from it.

Another implementation instrument mentioned explicitly in the Physical Planning Act is the construction permit (Article 14). This covers works such as demolition, road surfacing, irrigation, and changes in the ground profile. Such a permit is required only in an area covered by a land-use plan and only if that particular plan explicitly requires it. A construction permit may only be refused if the works would be in conflict with the land-use plan, or if they would conflict with the Monuments Act or a provincial or municipal ordinance.

When a municipality adopts a land-use plan that includes land been designated for industry, it should take account of the environmental effects of that industry. It can do this by:

○ stating the types of industry that will (or will not) be allowed (*staat van inrichting*: list of industries);

58

○ designating different areas of the land which has been designated industrial for different types of industry;

○ designating the land so that environmentally damaging uses are separated from uses sensitive to such damage.

The Public Nuisance Act (Article 3, Part 2) allows municipal councils to establish zoning regulations for part or all of their municipal areas. These can be applied in combination with land-use plans to reduce environmental nuisance from and within industrial areas (Neuerburg & Verfaille 1985).

Compulsory purchase can be used to implement a land-use plan. Compulsory purchase is possible using a clause in the Constitution itself, but only by preparing special legislation for each case separately. So a special Compulsory Purchase Act (Onteigeningwet) has been passed (the latest revisions took place in 1981). There are several grounds for compulsory purchase, among them "in the interest of physical planning and housing" (Article 77). Under this article, compulsory purchase may take place:

○ for implementing a land-use plan or for maintaining the existing situation where this is in conformity with the land-use plan;

○ for implementing a building plan (i.e. in connection with a building permit);

○ for clearing sites to be used for housing;

○ for clearing dwellings vacated or declared unfit for habitation.

Most compulsory purchase takes place under the first two grounds, especially in combination with a legally valid land-use plan.

An application for compulsory purchase is tested against its importance for physical planning and housing, the public interest, its necessity, its urgency and its financial feasibility. The test of public interest is very important. A municipality (as the plan implementor) must show that it has tried to purchase amicably at market price and that the present owner is incapable himself of implementing the plan (or that part of it on his own plot). The latter test can be very important. Municipalities that apply to acquire land compulsorily can be refused if the owner can claim convincingly that he is capable (financially and technically) of ensuring that his plot will be developed precisely in conformity with the land-use plan.

Compensation is regulated by Article 40. The principle is that the person whose property is purchased compulsorily must be fully compensated for all damages suffered directly and necessarily as a result of the loss of his property. This includes the price for the property but also all other damages, such as removal costs and the reduction in value of other goods not compulsorily purchased.

The price paid for the property is specified as: "the price that would arise by an assumed transaction in free negotiations between the compulsory purchaser and the compulsorily purchased, both acting as reasonable people"

(Article 40b). This is not the existing use value, nor market value, but the value in the use for which the land is being purchased, determined in an "objective" way (de Haan 1981). Furthermore, although the new land use is allowed to influence the price, the necessary public works may not (Article 40c). The result is what the legislation calls the "actual value" of land. These formulations are all very vague: to see how they have been interpreted one must study the case law. All that needs to be said here is that the compulsory purchase price lies somewhat above existing use value.

Compulsory purchase procedure is divided into two parts: administrative and legal. In the administrative part, the Crown is asked to approve the application for compulsory purchase, after which the case goes on to the second part, in which the court, accepting the approval of the application as being also an approval of the legal side of compulsory purchase, goes on to determine the amount of the compensation.

The second part lasts six months or more. But it is possible to ask the court to make an early judgement, whereby the applicant for the compulsory purchase is given disposal over the property on payment of 90% of the price that it has already offered. This speeded-up procedure is the one normally followed, and under it the purchasing party can get disposal two or three months after the decision of the Crown. In the meantime, the court continues to determine the exact amount of the compensation. Recently, the independence of the Crown as a court of justice has been questioned, as a result of which this procedure is being changed. One change is that the court would not accept the approval of the Crown as being a test of the legal merits of the compulsory purchase, but would make its own legal test. Already this is happening voluntarily in some cases.

Within an area specifically designated under the Town and Village Renewal Act for that purpose in areas covered by structure and land-use plans, the municipality has the right of first refusal on land offered for sale. The aim of this is to prevent land speculation without having to resort to compulsory purchase, which in any case takes too long to damp down speculation. The seller of land must inform the municipality, which has two months in which to exercise its option. The price is determined by negotiation or in the courts.

A municipality can impose charges on the developer of land, to contribute towards the costs of making the plan and of the necessary infrastructural works (the *bouwgrondbelasting*, discussed in Ch 3.3). This can be done when the landowner himself wishes to implement a plan (or that part of it on his own plot). Then, the municipal council can make its co-operation contingent on conditions laid down in its exploitation ordinance. This stipulates also which land is to be ceded to the municipality for public purposes.

The public powers available for reallocating landownership in rural areas (see Ch. 1.1) are not available in urban areas, although they have often been suggested (see, for example, de Haan 1981). They were created for use in an exceptional case: the rebuilding of the bombed centre of Rotterdam after the Second World War.

Instruments for implementing plans under private law

If a municipality itself is the owner of the land within a land-use plan, then it may use its position as landowner to help implement the plan. The powers this position gives have been described above. They can be used as implementation instruments as follows:

○ as supplier of land on which development is to take place, a municipality can choose to whom it supplies, when it supplies and in what sequence, and at what price;

○ using the powers to impose restrictive covenants and positive obligations, a municipality can specify what is to be built on the land, when building works must start and be finished, even the price at which the completed development must be sold or rented;

○ if a municipality decides to dispose of the land under ground leases instead of selling it, it can exercise some control over the use and maintenance of the development for the whole duration of the lease;

○ it has been suggested that a municipality could use its powers to impose "chain-agreements" when selling land in order to exercise long-term control over the development, similar to the long-term control possible with ground leases. This, however, has not been tested in law.

The combination of public law and private law

A municipality may always implement a land-use plan using its powers under public law. If a municipality is legal owner of the relevant land, it may use private powers. It may, in the latter case, use public and private powers in combination. It is generally agreed that a public body may realize its public responsibilities using private powers. "Only when public law makes it clear that a matter is to be realized exclusively by public powers may private powers not be applied" (de Goede et al. 1956). This restriction on using private powers is not specified by public law for physical planning, housing, or the environment.

This is an example of the move in legal thinking towards what is called the mixed doctrine of public and private law. Agencies of public administration use both types of law. Neither is superior to the other, but the agency is bound when applying both types to the general principles of responsible administration (*behoorlijk bestuur*) (Terpstra 1991).

Information systems

The land register and the cadaster Information on land and property is held systematically in various places, mainly for registration and for cartographic purposes. One of these places is the cadastral service (Dienst Kadaster en Openbare Registers). This is an agency of the ministry of housing, physical planning and environment. Its main tasks are:
○ keeping and updating the public register (the land registry function);
○ the system of triangulation;
○ contributing information to land consolidation projects (Ch 1.4);
○ producing the large-scale basic map of the Netherlands;
○ providing information to support the minister of housing, physical planning and environment.

The cadastral service includes:
○ the public register. Every transaction in land and property has to be established by act of a public notary (Ch. 1.1). The ownership rights are not legally transferred until the act has been entered into the public register;
○ the cadastral register, including the precise location and delineation of the property (*legger*), the name by which the property is to be referred to, and all the titles and deeds (*perceelsregister*).

These registers are automated and contain all information about the real (material) rights attached to property units.

The other most important place where information on land and property is held is the municipalities. These are obliged by law to maintain two real estate registers, the municipal cadaster and the dwelling register (which latter is part of the municipal register of its inhabitants). Many municipalities are automating their registers and digitalizing all property units. Then they should be able to link into the already automated national cadastral register.

Other public agencies that maintain registers of land and property include the water boards, public utilities, the provinces, and central government agencies of the ministries of finance, traffic and water management; agriculture and fisheries; housing town planning and environment; defence; and economic affairs.

Since 1984 there has been an advisory council for real estate information (Raad voor Vastgoedinformatie, RAVI). Its task is to advise the minister of housing, physical planning and environment (as co-ordinating minister) about all aspects of real estate information. Advice has so far been restricted to registration and cartography (RAVI 1988).

Other sources of information The national physical planning agency (Rijksplanologische Dienst) commissions every year the economic and technological institutes (ETIs) in the provinces to carry out surveys of industrial land. The aim to get a picture of changes in land availability on industrial estates. It is limited to estates where, at the beginning of the year, at least 5ha are available for use. The ETIs obtain the information from the municipalities. The findings are published annually in the private journal *VastGoedMarkt* (VGM). Information is given about the amount of land disposed of in the previous year, how much is immediately available (serviced), and how much is available but not yet serviced. Information is also collected about disposal prices and whether land is for sale, rent, or ground lease. Information is provided for each industrial estate separately, according to municipality and province, and is analyzed in Chapter 5.2.

The central bureau of statistics (CBS) used to collect and publish information about transactions in building land. The results were published in *Maandstatistiek Bouwnijverheid* and included the number of transactions, average sales price, area sold, whether sold for use by industry or for other uses, and whether sold by a municipality or another authority. Transactions for other uses, such as recreation, highways, public open space and car parking were not included. It was based on a sample survey, and the results were statistically reliable. Figures were published over the period 1965–83. Then the survey ceased, to reduce public expenditure. The figures up to 1984 are analyzed in Chapter 5.2.

A private consultancy, TAUW Infra-Consult, has for several years now been commissioned by the ministry of housing, physical planning and environment to conduct a survey of building land: plot sizes, plot prices, and prices per square metre. The figures are analyzed by land use (single family housing, multi-family housing, other uses), and subdivided further by type of housing. Information is collected from 79 municipalities, and the results give a statistically reliable picture of prices for building land and of trends in the amount of land disposed of for non-housing uses. The survey is restricted to greenfield developments: urban renewal areas are not included. The results are analyzed in Chapter 5.3.

3.2 The financial environment

The financing of property development.
Here we must make a distinction between the public sector (some housing, road building, water management) and the private sector. Most building in the public sector is financed out of central government income. In addition, funds are sometimes borrowed (especially by municipalities) from banks. Private property development is financed by institutional investors or by bank loans or out of the developer's reserves.

The financing of property ownership

The user of a plot of land can own it outright, financing the purchase either from capital or with a loan (usually in the form of a mortgage). Or the user can own a lease on the land, paying the lease either periodically or as a lump sum (a premium); finance for a premium can be borrowed by means of a mortgage.

A mortgage bank (*hypotheekbank*) or an ordinary commercial bank is often used to finance real estate purchases. Mortgage banks usually limit their involvement to a percentage (eg 70%) of the assessed value of the property. The remainder is often financed by other sources, for example from reserves or, for commercial property, by a loan from a commercial bank secured by a guarantee from the parent company of the borrower and secured by a second mortgage.

Banks usually charge a commission (*afsluitprovisie*) of 1–2%. Interest is usually charged monthly. Some mortgage banks require the mortgagor to pay interest monthly in advance. The repayment can take many forms. The mortgage is vested by a notarial deed that is usually executed simultaneously with the loan agreement and the conveyancing deed. A copy of the mortgage deed must be filed with the land register office. The bank remits the funds to the notary, who is permitted to distribute the funds after the mortgage deed has been registered in the property register. The notary charges his fee on the basis of the notarial rates, relating to the total amount of the secured debt.

The financing of property use

If the user is also the owner, property use is financed as property ownership, which is described above. If the user rents the property, the rent is paid out of personal or operating income. There are, however, a number of possibilities that can be regarded as intermediate between ownership and renting. Two of them will be explained here: leasing and *huurkoop*. As Chapter 1.1 pointed out, tenancy rights cannot be transferred. Chapter 7.1 explains that most tenancies are short-term, for five years. Leasing can be regarded as an attempt to fulfil the need for security in long-term tenancies.

There are two forms of leasing: finance lease and operational lease. With a finance lease the lessee acquires economic ownership at the beginning of the lease, while the lessor retains legal ownership during the lease but subject to the lessee using his option to acquire the legal ownership at intervals of, say, every five years. At the expiration of the lease, the lessee has the right to acquire the legal ownership for a nominal sum. The lessee cannot terminate this contract during its course. The risk during the lease period is borne by the lessee. The lessor has in fact only the function of financing the object. (See Ch. 1.1 for an explanation of the difference between legal and economic ownership).

With an operational lease, the lessor retains not only the legal ownership but also the economic ownership, even after expiration of the lease contract. However, the lessee has the option to buy during the period of lease.

The finance lease is quite commonly used. For the user, it has the advantages that it makes no claim on capital or credit: outgoings are fixed and cannot fluctuate. And the user can determine at the beginning of the lease what he will do (buy or vacate) at the end of it. Furthermore, there is no liability for transfer tax (see Ch. 3.3) when economic ownership is transferred at the beginning of the lease, only if legal ownership is transferred later. However, legislation passed to protect lessees of dwellings (Tijdelijke wet huur-koop onroerend goed, 1973) has been interpreted as also applying to financial leases; and this legislation provides that the lessee can terminate the lease. That is not attractive to the lessor; if he has financed the lease when interest rates were high and rates then fall, the lessee might want to terminate the contract, leaving the lessor with a fixed-interest obligation that he cannot cover at the lower interest rates. So financial leases are often not attractive to lessors.

With *huurkoop* (literally, hire purchase) the user pays a sum periodically for a number of years to the legal owner under a contract whereby the user becomes the legal owner when the pre-arranged payments have been completed. It is a sort of deferred purchase. As a method by which the user finances his occupation of real estate this is not very common.

Restrictions on capital imports and exports
In the Netherlands there are no restrictions on capital imports and exports. Real estate developers and funds are free to make real estate investments abroad. There is, however, a restriction on one of the largest institutional investors, the Algemeen Burgelijke Pensioenfonds, which is the pension fund for all who work in the public service. This may invest no more than 5% of its total investment outside the country. Also real estate investors from abroad are free to invest in capital assets in the Netherlands.

The financing of land acquisition and development
Land policy is regarded mainly as an instrument of physical planning at the local scale. For this reason, Dutch municipalities provide much of the building land. This requires huge outlays of money that the municipalities hope to recoup by land disposals. Their investments are largely financed by bank loans, secured on the prospects of serviced building land. This implies risks for the municipalities: when land disposals stagnate because of shortage of demand, the municipalities are confronted with losses on their investments (see also Chs 1.4 and 4.1). The financing of property development,

ownership and use per sector is discussed in Chapters 7.2 and 8.2.

Land banking

Land banking is not common. As Chapter 4.1 indicates, most building land is supplied by municipalities. These acquire the necessary land for each land-use plan area separately: they do not experience much difficulty with this, nor do their activities push up prices significantly, so they have no need to build up land banks in advance. And property developers have little need to build up land banks if the municipality is a reliable supplier of land.

Land transaction costs

Taxes (*overdrachtsbelasting*) are levied on the transfer of legal ownership of, and real or material rights in, landed property. This transfer (or convey-ancing) tax is regulated in the Wet op belastingen van rechtsverkeer, 1970. The tax level is based on the value of the property and is set at 6%. This applies also to the establishment of a real right such as a ground lease, but not to transfers of economic ownership (as in financial leases). The tax is levied on the new owner. Transfers have to be established by an act of a public notary and registered (see Ch. 1.1). The tax has to be paid at regis-tration, and the notary is personally responsible for this. If a transfer takes place within three months of a previous transfer, no tax has to be paid (some property speculators take advantage of this provision).

When property is transferred, the services of a notary must be engaged, for which a fee must be paid (1.5% of the value, paid by the purchaser). If an estate agent is employed, the fee is 1.85% of the value. If the property is valued, the valuer charges 0.185% of the capital value. Value-added tax (*belasting op toegevoegde waarde:* BTW) has to be paid on all these fees. If a mortgage is required, then the costs of arranging this must also be paid.

The tax on turnover (BTW) is also of great importance for the land and property markets. This is an indirect and general excise tax on all consumer expenditure. It is levied on all businesses in the chain of production: manufacturer, wholesaler, and retailer. The tariff is 18.5%, 6% or 0% of the purchase price net of tax, irrespective of the number of links in the chain. Every link pays the tax on the balance of the value it adds, hence the name value-added tax. Every business charges the tax to its customer. If the customer is also a business, it may deduct the BTW it has already paid. Because the final consumer has no right to deduct previous taxes, he pays the whole of the tax.

Value-added tax is applied when landed property is obtained, at a tariff of 18.5% and according to the following rules:

○ the concept of "obtaining" includes both the transfer of legal and of economic ownership (and, therefore, to financial leases, but not to

operational leases);
○ in general, "obtaining" landed property is exempted from BTW. There are two exceptions: (a) obtaining newly built property before, within, or up to two years after the first occupation; (b) if the buyer and seller jointly apply to pay the BTW, in which case they are exempted from the transfer tax;
○ it follows from this general rule that obtaining undeveloped land is exempted (but not exempted from the transfer tax), but that the servicing of land is subject to BTW (but not to the transfer tax);
○ new property that is not yet occupied attracts BTW: in that case, transfer tax does not have to be paid;
○ the "obtaining" of property for rent is not subject to the tax.

3.3 The tax and subsidy environment

Taxes concerning the land market

Taxation of capital assets In the Netherlands a net wealth tax of 8% per annum is levied. Individuals both resident and non-resident are subject to this tax. Non-residents are only subject to tax concerning certain elements (such as real estate) of their net wealth located in the Netherlands. If the real estate is sufficiently leveraged this tax can be avoided or minimized.

Taxation of landed property An annual property tax (*onroerend-goedbelasting*) is levied by the municipality where the real estate is situated. Both corporations (resident and non-resident) and individuals are subject to it. The tax is levied on the basis of the fair market value of the real estate, free of encumbrances and available for sale (Gemeentewet, Article 273). Valuations for this tax are described in Chapter 8.1.

The taxable object is the property as it is entered in the cadastral register (see Ch. 3.1). All objects within the municipal boundary are subject to the tax, with the following obligatory exemptions: agricultural land, churches, nature reserves and objects of natural beauty, public rights of way, railway lines and waterways, works for water defence and water management, and water purification works. In addition, the municipality may, but is not obliged to, exempt the following: municipal buildings; street furniture; cemeteries; public open space; facilities for sport, education, and health; and dwellings below a certain value.

The tax is levied separately on owners and users. The tariff is set by the municipality, and tariffs for owner and users are different, although the difference must not exceed specified limits. Moreover, the total amount raised by a municipality from its property tax is subject to specific limits. Within these, there is a wide variety in the tax burden between municipalities. The municipality of Nijmegen is not untypical. In 1990, the rates were: for every complete 3,000Hfl of value, 8.15Hfl to be paid by the owner, 6.55Hfl to be paid by the user: together, therefore, a property tax of 0.5%.

Taxation of income and capital gains Income tax is levied on companies as corporate income tax, and on individuals as personal income tax. Income received from rent is always liable to tax, whether from property held for business purposes (company income tax) or as a personal possession (personal income tax).

Under company income tax, a user of a rented building can offset rent against tax. A user of his own building, but also a lessee of a financial lease, can deduct against tax interest payment on loans for the property, as well as the depreciation of the building. Ground rents paid under a ground lease are also tax deductible. Under personal income tax, only interest payments on loans for the property and ground rents are tax deductible.

In 1991, the maximum rate for personal income tax was 60% (starting at a taxable income of approximately 90,000Hfl). Corporate income tax rates were 40% on the first 250,000Hfl of taxable income and 35% on the remainder. There are no special tax rates for capital gains: such gains are regarded as income and are subject to tax at the ordinary rates (unless the proceeds are deposited in a replacement reserve, where they can stay for up to four years). However, capital gains realized from the sale of real estate by individuals (personal income tax) are not subject to income tax, unless such income is derived through an individual's trade or business and the real estate was part of the business assets of the individual. It will be evident that where capital gains are not taxed, capital losses cannot be regarded as tax deductible.

Pension funds are regarded as companies that have passive investments in real estate as their purpose. These companies benefit from a zero corporate tax rate, but are obliged to distribute their earnings to their shareholders, who are then taxed on the dividends thus received.

An international investor may own Netherlands real estate directly or indirectly through Dutch entities or through non-resident entities. Such an entity may be a corporation or a partnership. These alternatives have different tax consequences; investments via a partnership are generally more in line with direct investments rather than indirect investments.

An individual Dutch resident is taxed on his worldwide income (income tax on, for example, investment income, business income and employment income). Credits are allowed according to the various tax treaties. Non-resident individuals are taxable for income derived from specific sources, such as business income and income from real estate located in the Netherlands.

Taxation of betterment If landed property benefits from municipal provisions (e.g. public works such as new roads), a tax (*baatbelasting*) can be levied on the owners of real rights in that property, but not on tenants (see Ch. 1.1), as a contribution to the costs incurred. The tax is levied annually for not longer than 30 years and is regulated in the Gemeentewet, Article 273a. It would appear that this regulation allows taxation of betterment in the classical sense (betterment IIa, see Kruijt et al. 1990: 33). In practice, however, it is rarely applied: practitioners say that it is almost impossible to levy.

Taxation of building land If land becomes better suited for building upon as a result of municipal provisions (e.g. land drainage), a tax can be levied on the land (*bouwgrondbelasting*) as a contribution to the costs incurred. The tax is levied annually for not longer than 30 years. When determining its level, a designation (land-use) for the building land has to be specified. That is regulated in the Gemeentewet, Article 274.

Chapter 4.1 shows how most building land is supplied by municipalities. When determining the disposal price, the municipality aims to cover its costs for infrastructure works, etc. In those cases, it is not necessary to levy the *bouwgrondbelasting* in order to cover costs; moreover, the legislation specifically excludes the use of the tax when the costs are covered by land sales, etc. The significance of the tax on building land is that it can be used to oblige landowners to contribute to the costs of necessary public works when the land has not been taken into municipal ownership. However, case law has determined that a contribution can be exacted only for public works that directly benefit the landowner.

Taxation of gifts and inherited property A tax (*successierecht*) is levied on the "obtaining" of gifts and inherited property when the person who gives or who leaves the inheritance is a resident of the Netherlands. The beneficiary is liable to the tax. The tariff levied is dependent on the relationship of the beneficiary to the donor or inheritee, and on the value of the benefit, with exemption up to a specified value.. For example, a spouse may inherit almost 500,000Hfl before having to pay the tax. Landed property received in this way is subject to the tax, which is regulated by the Successiewet, 1985.

Subsidies for the land development process

By the land development process we mean the acquisition of land that is to be developed or redeveloped, its servicing (or re-servicing) so that it can be built upon in accordance with the plan for (re-) development, and the disposal of the serviced land to the building developer. (The building developer may be the same person as the land developer, but it is helpful to distinguish between the two processes: land development and building development.) When a municipality is the land developer, central government makes available two sorts of land development subsidy. These subsidies are granted only in special circumstances, as described below.

The *lokatiesubsidie* is given under the following circumstances. When a municipality develops land, it usually disposes of some of it for socially rented housing. The resulting price per plot, or the total price of the dwelling, has to fall within certain norms or guidelines set by central government (which, after all, provides the housing subsidies, see Ch. 7.3). It can happen that the disposal price of such land is too high, according to those norms or guidelines, and that there is nothing a municipality can do to bring prices lower. This is often because of unusually high land-servicing costs (see Ch. 4.1). This could be a reason for deciding not to develop the land and choosing a different location. If, however, central government considers that there are overwhelming reasons for developing on a high-cost location, it will subsidize the costs until plot prices for socially rented housing have fallen to the norm (guideline) level.

The second type of central government subsidy is called the "large development area subsidy" (*subsidie grote bouwlokaties*). This is granted only occasionally when a development is in the national interest. Examples include redevelopments along the River IJ in central Amsterdam and the redevelopment of the site of the Sfinx-Céramique factory in the centre of Maastricht. For greenfield development, this subsidy is additional to the *lokatiesubsidie*, for urban renewal it is additional to the urban renewal fund (see below). In such exceptional cases, central government might find yet other, additional ways of subsidizing a project, and the province might also provide a subsidy. From 1 January 1994, the *lokatiesubsidie* and the *subsidie grote bouwlokaties* will be amalgamated into the *Besluit lokatiegebonden* subsidies (location-specific subsidies).

Until 1984, there was a central government grant towards the land redevelopment process in urban renewal areas: this was one of more than 40 separate central government grants for urban renewal. When the Town and Village Renewal Act was passed in 1985, most such grants were amalgamated into one budget, the urban renewal fund. This is distributed between the municipalities (see also Ch. 2.1), who are free to decide how to use it, as long as it is for urban renewal. Some municipalities will use it for subsidizing the land redevelopment process. Note that the use of this fund is not restricted to areas covered by an urban renewal plan.

Where soil is contaminated, subsidies are available towards the costs of cleansing it. The procedure is as follows. The province makes a five-year plan for decontaminating ground and is responsible for implementing it. Costs are paid partly by central government, partly by the province, and partly by the municipality where the contamination is found. However, these public contributions can be partly recouped from those who were responsible for the contamination. This is regulated in the Interim Act on Soil Decontamination (Interim Wet Bodemsanering, see Ch. 3.1).

Where land development costs would be especially high because of the need to build screens to insulate housing and other amenities from noise caused by roads, railways, etc. (see discussion of the Noise Nuisance Act in Ch. 3.1) a subsidy can be paid by central government towards the cost of the works (*Bijdrage regeling geluidwerende maatregelen*), including the costs of fitting the dwellings with a noise-insulating facade (*geluidswerende gevelvoorzieningen*).

A municipality may decide to provide subsidies from its own income or reserves for the land (re-)development process. And within one land development area (one plan area), a municipality is free to set disposal prices so that there is what could be regarded as cross-subsidization, from high-paying uses to low-paying ones. Both these forms of subsidy are described in Chapter 4.1, where the very general check exercised by provinces on this form of subsidy is also described.

When the land developer is private, no subsidies are available. Under tax law, interest charges on loans are tax deductible in all cases, including loans for private land development. But depreciation of land as a capital asset is not tax deductible; the principle is that land does not depreciate (in contrast to buildings).

Property subsidies for new buildings, which affect the land market by increasing the demand for land, are described in Chapter 7.3.

Compensation for loss of value

Compensation is payable for material damage caused by the actions of the planning authority under certain conditions. This is regulated by Article 49 of the Physical Planning Act 1962 (modified in 1985) which states:

> If and to the extent that someone with a material interest suffers damages as a result of:
> o   the content of the land-use plan (bestemmingsplan);
> o   a number of other actions of the planning authority;
> and it is not reasonable to expect that person to accept the damage and if the damage is not compensated by purchase or compulsory purchase, then the municipal council grants, at his request, fair compensation.

In the introduction to this legislation it says:

Damage caused by land-use plans is damage caused by restrictions on the freedom of individual citizens. But the public administration is not required to pay compensation for measures which go no further than making concrete the limitations on freedom which arise when a citizen shares a restricted territory with others. It can be, however, that a measure is taken which restricts the freedom of a citizen more than is general for the social situation. Only in that case is compensation payable.

The civil code (Book 6, Article 162) also regulates compensation, but in this case under private law (i.e. damage caused by one private citizen to another). It is not applicable to the actions of the planning authority acting as a public body; but certain general principles from private law are deemed to be applicable to public law.

The legislation and the case law that has grown up around it has led to the following conclusions:

○ the damage must have been caused by a change in the content of the land-use plan. If a plan is made for the particular area for the first time, then the change is from "no plan" to "plan", which is not regarded as being a change in the content of the land-use plan. In other words, the damage caused by the introduction of land-use planning is not liable to compensation;

○ the damage must have been caused by the actions of the planning authority. Moreover, when determining the extent of the compensation, the principle is applied that every citizen must expect some disadvantage from actions of the public administration;

○ it is a principle of the law on private compensation that no citizen can expect to live without risks and that, therefore, not all material damage can be laid at the door of another (Civil Code, Book 6, Article 101). This principle can reduce the extent of the compensation payable. For example, suppose someone owns land on which, under the plan, offices can be built. If the plan is changed so that only houses may be built there, compensation should theoretically be payable. But if the landowner had taken no action to develop the land in the 10 years before the change he has clearly run the risk of the plan being changed. It is therefore considered that it is his own fault that he has missed the boat, and the compensation payable may be nil.

From these conclusions it will be obvious that this sort of compensation is a "non-issue". Compensation is rarely paid, and the existence of the regulation has little effect on the content of a land-use plan.

CHAPTER 4
The process

4.1 Price setting

Land development by the municipalities

In order to describe and explain price-setting on the urban land market it is necessary to anticipate Ch 4.2 and 4.3. The point is that most land development and redevelopment is carried out by municipalities. Most of the land on which new building takes place is supplied by municipalities (figures are given in Ch. 5.2). Much of the other land on which new industrial building takes place is supplied by harbour boards, which are municipal agencies carrying out land development much as a municipality would. And much of the remaining land is, in all probability, infill sites. The result is that price setting is dominated by the way in which municipalities develop land. Even in those occasional cases when the land development process is in private hands, market considerations (the private land developers are, after all, competing against the municipalities) will oblige private developers to follow the prices set by public developers.

The process of land development by municipalities can be described briefly as follows (some parts of the process are described more fully later). Usually the municipalities do not hold "banks" of land suitable for building; and they acquire the land for each development scheme separately. A land-use plan is made and within the designated area the land is purchased. If it is a greenfield site, usually all the land is bought; if the site is already built upon it might be too difficult to acquire everything, but arrangements can be made with owners of the properties not acquired concerning improvements and a contribution to the infrastructure costs (*bouwgrondbelasting*, see Ch. 3.3).

At the next stage the municipality services the land within the designated area, laying out gas, water, electricity and drainage, and building roads, car parking areas, public open space, etc. Often the land is so wet that it must be drained, or the level must be raised before any further works can begin. This is exceedingly expensive.

A large part of the land will remain in municipal ownership: the part used for roads, paths, play spaces, planting, etc. The rest is to be disposed of, mostly as freehold, sometimes leasehold (the choice having little effect on the disposal price). Disposals are to developers, which include developers of commercial and industrial buildings, private housing developers, housing associations building for sale and rent, semi-public trusts for schools, old people's homes, hospitals, sports clubs, and so on.

The recommended practice for setting disposal prices is that, for each project as a whole (everything within the designated area of the land-use plan), the income from disposals has to be at least sufficient to cover costs. These latter include costs of acquisition, compensation, laying infrastructure and services, and the costs of preparing the plan, professional fees, and interest charges. The total income from disposals must be sufficient to cover costs, but how is that income to be raised from the various plots that are to be disposed of?

First, the price for commercial uses is decided; either a market value or some multiple of the housing value (the determination of which is the last stage in this calculation). Second, the prices at which non-commercial uses such as playing fields, schools, etc, are to be transferred to municipal departments, school boards, etc. is decided, either based on a norm (so much per square metre) or related to the housing value. The income from these sources is subtracted from the total (minimum) income that has to be raised; what remains has to come from disposals for housing. All the housing plots are then given weights, depending on the type of housing (social rented, subsidized owner-occupation, unsubsidized), the location (with or without a good view, etc.), and the plot size. The income to be raised is divided between the plots according to these weights, hence the disposal prices for housing. Until 1990, the plot prices for the heavily subsidized rented dwellings (*woningwetwoningen*) had to be approved by a central government agency. Since 1990, central government has restricted itself to publishing guidelines for land prices (except in the "*lokatiesubsidie*" areas, see Ch. 3.3, where maximum plot prices are imposed).

The key to understanding the price setting is to be found in the financial statement (*exploitatierekening*) that accompanies all land-use plans. This statement is not required explicitly by legislation. However, plan preparation must be accompanied by research into the feasibility of the plan, including its financial feasibility (see Ch. 3.1). The results of the research are appended to the formal plan documents (map, regulations, etc.) as explanatory notes. And although these notes are not legally binding, they are treated very seriously by the province when it decides whether to approve the land-use plan. So the financial statement is always scrutinized carefully. Moreover, in so far as plan implementation will require central government subsidies(e.g. for housing), the decision to grant such subsidies will usually be made dependent on the financial statement being found satisfactory.

74

The plan area as a whole may be treated as the accounting unit for the financial statement, or for convenience it may be divided into a few areas, each with its own financial statement. Each financial statement must show how the costs incurred by the municipality are to be covered and so it has a cost-side and an income side. Costs include the costs of making the plan itself (either by the municipality's own staff or consultants commissioned by it). These can be covered by levying a standard charge of between 19% and 22% on total land servicing costs (Prins, 1985).

Whenever a municipality has acquired the land on which development is to take place (the vast majority of cases) then the following costs have to be added: costs of acquiring the land and paying compensation; costs of servicing the land (these will usually include the costs of draining the land, sometimes even of raising its level); putting in gas, water and electricity, foul and surface water drains, roads, footpaths, cycle tracks, parking spaces, street lighting, etc; and interest charges on capital costs until these are recouped.

A large-scale development very often requires new infrastructure that benefits the whole town. In these cases, a part of the costs of these supra-district facilities (*bovenwijkse voorzieningen*) is added. Who is to pay the remainder of these costs is a matter for negotiation between municipality, province and central government. For the part that is to be recovered from the plan itself, there are several possibilities (see de Graaf 1984, para 1. 7). For example, a charge of 10% of total land and servicing costs or 3Hfl per square metre of land disposed of may be levied. In that way a municipality can build up an urban development fund and differences in such costs between various large-scale developments can be evened out.

The income can be of three sorts: from levies, land disposals and subsidies. If the municipality were to restrict itself to making the plan and issuing permits (i.e. taking no direct involvement in the land), its own costs would be low and could be recouped by levying charges on developers (under the taxation of building land, see Ch. 3.3). Where the municipality acts as the land developer, which is what usually happens, it can raise income by disposing of the land either freehold or by ground leases. If these two sources of income are inadequate, then there must be subsidies.

The financial statement must show how costs are to be covered. Losses are not acceptable, profits are not forbidden, but the expected aim is to do no more than cover costs. It must be added that these were the common practices, and municipalities are now rather freer in the way they carry out land development. However, most post-war building has been shaped by these "common practices", and the way of thinking that has grown up around them will not change quickly.

Because most development and redevelopment takes place on land supplied by municipalities, the costs of the land development process are borne by them, and they enjoy the returns. Municipalities may therefore make profits, but they may also suffer losses. Most building development, on the other hand, is done privately or by semi-public trusts (e.g. housing associations). This division of labour brings with it a certain division of costs, returns and risks in any scheme where land and buildings are developed. Increasingly, attention is being given to arrangements for a different division of labour that would divide costs, returns and risks differently. These go under the name of public–private partnerships. One of the possibilities is that land development be done by private bodies. However, the present system is so dominant and so widely accepted that it would take many years and radical changes before the prices to be described below were greatly affected.

Acquisition of land for (re-)development

It is common practice for a municipality to acquire all the land within the boundaries of its plan area. This is financed by loans arranged privately and secured on the building land to be disposed of. Acquisition can be amicable or by compulsory purchase: proceedings for the latter cannot start until amicable acquisition has been tried and has failed. Amicable acquisition can start before the plan has achieved legal status, and usually does. It is only when acquisition has to be compulsory that the municipality has to wait until the land-use plan is legally valid, for that gives the usual basis for compulsory purchase. However, it is striking how rarely compulsory purchase powers are used by municipalities: between 1979 and 1982, only 0.06% of the total area acquired by municipalities came under this category (CBS, Maandstatistiek Bouwnijverheid).

Amicable acquisition has to take place at market prices, at the least; otherwise the owner would not sell amicably. But how can a meaning be given to market prices in a situation where, within one municipal area, a municipality itself is the only demander, and therefore a monopsonist of land to be serviced for building? On greenfield sites, existing use value is the minimum that may be offered (agricultural land values are given in Ch. 5.3). In urban areas (redevelopment), existing use value is again the least that may be offered. But will not the existing owners hold out for full development value, if necessary forcing the municipality to go to compulsory purchase?

In this monopsonistic situation, the way in which legislation fixes compulsory purchase prices is crucial. For existing owners will hold out for compulsory purchase only if they gain by it. However, we have seen that the law on compulsory purchase is anything but clear on this point. What happens in practice?

The situation is resolved to most people's satisfaction in the following way. The practice has grown up that compulsory purchase price is set a little above existing use value. Municipalities have also made it the practice to offer to acquire amicably but at compulsory purchase price levels. And most owners will sell amicably because they receive more than existing use value without the delay of compulsory purchase. And the municipality is happy because the costs and delay of compulsory purchase have been avoided. So we can explain the paradox that compulsory purchase is hardly ever used and yet its sanction is essential if the municipality is to acquire land.

There are exceptions to this happy, if ad hoc practice. A property developer might buy from an existing land owner within a designated plan area, anticipating the municipality and possibly offering a higher price. The developer may then refuse to sell to the municipality, which cannot use compulsory purchase powers if the developer can show that he is able to provide the development exactly as shown in the plan (see Ch. 3.1). The developer might do this if the gains to him would be higher than the compulsory purchase price offered by the municipality. These gains will be greater if owner-occupied housing is to be built on the land owned by the developer. Some developers play this tactic for other reasons, namely to acquire a stronger bargaining position for winning building contracts from the municipality. If this happens, the municipality can recoup some of the infrastructure costs it has made through taxation (see Ch. 3.3) and the plan will still be implemented more or less as intended. There is no evidence that it is a serious obstacle to the desired functioning of the land market.

If agreement cannot be amicably reached between municipality and existing owner, then many municipalities will offer to go to arbitration (see below) rather than start compulsory purchase proceedings.

Disposal of building land

After the municipality has acquired the land, it services it in accordance with the plan. The costs of this are also financed by private loans, secured on the value of the building land to be created. Then the municipality has a number of building plots to dispose of.

The plots may be sold (legal ownership is transferred) or user rights may be sold (such as ground leases, building rights, or tenancies). In practice, only two disposal forms are important: sales and ground leases (*erfpacht*). However, the difference between sale and ground lease is not important for the disposal price. This is because the municipality determines the level of the ground lease by reference to what the price level would be were legal ownership sold. If the ground lessee pays an annual ground rent, this is more or less the same as the mortgage payments he would have paid had he bought the land. And if the ground lessee uses the option to commute the annual ground rents into a premium, this latter is more or less the same as the sale price. So the question can be simply asked: how are disposal prices set?

First, by reference to the total costs within the plan area as the accounting unit (see above). A municipality aims to ensure that total income from disposals at least covers total costs. In this way, an average disposal price per square metre of disposable land may be calculated, so that costs are covered. However, it is not the intention that all building plots be sold at the same price: some plots are to be used for housing and some for shops, some housing plots have a better location than other housing plots, etc. The way in which the municipality tries to divide the total costs to be recovered between the different building sites has been described above.

Because the total costs of the land development process affect disposal prices, it is important to say something more about how high these costs are. Figures have been published, but they have not been collected in such a way that they can be regarded as representative. We can, however, put together the following picture.

When municipalities prepared greenfield sites for building housing, their total costs (including acquisition) in the 1970s were around 50Hfl per square metre total plan area (SEO 1980); in later years a little higher (NIROV 1985). We restrict ourselves to the 1970s, for which years complementary information is available. When municipalities acquired land for urban development, they paid (excluding compensation for buildings and for other works on the land) about 10Hfl per square metre (CBS, Maandstatistiek bouwnijverheid). The price of agricultural land traded within the agricultural sector in the second half of the 1970s was between 2Hfl and 5Hfl per square metre (see Ch. 5.3).

Together, this would suggest the following picture for land development costs per square metre of total plan area for greenfield sites in the second half of the 1970s:

○ compensation for loss of existing use: 4Hfl;
○ compensation for loss of buildings, disturbance, plus gains to the supplier of land: 6Hfl;
○ costs of preparing the building sites: ≃40Hfl.

That makes clear the very high costs of servicing in relation to acquisition. Figures for greenfield development carried out by the municipality of Rotterdam support this conclusion: up to 90% of total land development costs was for servicing (de Graaf 1984). Since the second half of the 1970s those costs have risen, but there is no reason to think that they have changed greatly.

There will be much more variation in land development costs in urban renewal areas, so we cannot generalize from the scanty data available. Some figures available for Rotterdam (de Graaf 1984) confirm what we would expect: that acquisition costs are a much larger part of total costs (in Rotterdam, 55%) than on greenfield sites.

These costs are incurred when preparing the whole plan area. However, they are borne only by that part of it disposed of as building sites. That proportion is around 50% (SEO 1980, NIROV 1985). So the average costs per square metre total plan area ($=X$) give average costs per square metre disposable land that are twice as high ($=2X$). The starting point is that total costs should be recovered by disposals. So average disposal prices should be at least $2X$ per square metre. Disposal price for a particular plot will vary with land use, the location, etc.

However, there are limits to disposal prices. (If that were not so, land development costs could always be recovered and subsidies would never be necessary.) Limits are set as follows:

○ for plots to be disposed of for Housing Act dwellings, central government sets norm values for the plots (since 1990, only guidelines for those prices);

○ for other socially rented housing and for subsidized owner-occupied housing, central government sets norms for the all-in costs (land plus buildings);

○ for plots to be disposed of for marketable uses (private housing, shops, offices, etc), the property thus developed acquires a market value. From that value, the residual value of the plot can be estimated; a building developer will not be willing to pay more than the residual value. The market value of the developed property, and hence the residual value of the land, will depend on the volume of land supplied by the municipality (see Ch. 4.2).

In these ways, maximum disposal prices are set for most of the land to be disposed of. If a municipality finds that it cannot recover its costs without exceeding them, it can do one or more of the following:

○ try to increase income, by increasing the amount of disposable land within the plan area;

○ try to reduce total costs, by cheaper servicing, etc;

○ try to reallocate the costs between the plots to be disposed of;

○ try to find a subsidy.

If, on the other hand, a municipality finds that it can recover its total costs without exceeding maximum disposal prices, then it can do one or more of the following:

○ charge cost-covering prices, even though it could charge more;

○ raise the quality of the plan, by better infrastructure, more open space, lower density, etc;

○ raise disposal prices to the maximum, thus making a profit.

The land prices that result are presented in Chapter 5.3. An attempt to explain them more fully is to be found in Needham (1992).

Arbitration

The question of arbitration hardly arises when a municipality is disposing of land, for potential buyers can always withdraw if asking prices are too high. And even in the Dutch situation when there is only one supplier of building land (the municipality), the supplier is monopolist only within its boundaries. If a would-be buyer finds the price too high, he may go to another municipality. Cartels of municipalities to keep prices high are unknown and would in any case be inconsistent with municipalities' reasons for carrying out land development.

The situation is different when a municipality acquires land. For within a plan area, it is the only demander and the existing owner might find that he wants to sell (e.g. for personal reasons) but can find no other buyer except the municipality. Obviously the possibility of holding out for compulsory purchase strengthens the owner's position: but only if he can afford to wait.

That problem is most pressing in urban areas, whenever it has been announced that renewal is to take place but not for several years. Most municipalities where this occurs introduce arbitration procedures to protect the position of existing owners: the politicians on the municipal council would not accept the exploitation that might otherwise take place.

Where a municipality exercises its right of first refusal, the possibility of arbitration is included in the procedures. And when compulsory purchase powers are used, the legal procedure is in two stages (see Ch. 3.1). It is in the second stage that the compensation is determined. This is by a court, after taking the advice of a commission of experts.

4.2 The actors and their behaviour

The municipality

A municipality plays a double rôle on the urban land market: it is obliged, as the public authority for physical planning, to steer and regulate changes in the use of urban land; and it is able if it wishes to supply the necessary land for urban development and redevelopment.

These two rôles can be combined in different ways. For example, the municipality may choose not to operate actively on the land market; it may use its position as planning authority to strengthen its position on the land market, which it uses for purposes other than planning (eg financial); or it may take advantage of the possibility of active involvement on the land market to strengthen the implementation of its planning policy.

Most of the time most Dutch municipalities do the latter. In their rôle as planning authorities, by using their public powers they can guide urban development in their areas. However they want to lead as well as guide, so they complement their public powers with private powers, which they must first obtain by acquiring the land. It is not merely a case of municipalities choosing to do this: the provinces and central government expect them to do it and base much of their policy (e.g. for housing and for industry) upon that expectation.

If a municipality participates actively on the urban land market for that reason, how would it be expected to determine the price and the volume at which it supplies building land? And how do Dutch municipalities do this in practice?

How should a municipality set disposal prices in the pursuit of its planning policy? The simple answer is: as low as possible. For then the chances of attracting desired development and refusing undesired development are greatest. But the municipality must be able to cover its costs, or it will have to withdraw from active participation on the land market. So the answer becomes: as low as possible, as long as total disposals cover total costs.

That answer is adequate for non-marketable uses, such as social rented housing and playing fields. But land disposed of for marketable uses has a market value equal to its residual value. Following the rule "as cheap as possible" could produce disposal prices below that market value. This could lead to land speculation, which could endanger physical planning policy. That possibility can, however, be excluded by provisions in the sales contract (*anti-speculatie-bedingen*: if the land is resold within a certain period, a part of any increase in land value has to be paid to the municipality). This is sometimes done when land is supplied for subsidized housing for owner occupation.

Another objection to disposal prices lower than market value for marketable uses is that the purchaser of the land immediately acquires property with a value greater than he paid for it. Why should the developer enjoy that gain? Why not the municipality?

The rule when setting disposal prices should therefore be: as low as possible for non-marketable uses and market value for marketable uses. The way in which municipalities set disposal prices in practice has been described in Chapter 4.1. Does it follow the above rule? The recommended practice is: disposal prices as cheap as possible, as long as land development costs are covered (even if that means disposing to marketable uses below the maximum realizable price). But that is a dangerous practice because over the long period that the land development of a plan area takes, many things can change, resulting in expenditure that goes beyond the budget. In the first half of the 1980s some municipalities suffered enormous losses on land development (see Ch. 5.2). However, nowadays one hears little about such losses, more about the reserves that municipalities have built up on their "land account". Clearly, many municipalities set disposal prices that more than cover the costs of land development.

The second important aspect of the supply of building land by municipalities is the volume. In the land-use plan, building land is designated and the municipality also sees it as its responsibility to ensure that it is made available. There can therefore be no shortage of building land as a result of designated land not being brought onto the market. But could there be a shortage because insufficient land is designated? What is the behaviour of municipalities in this respect?

The general answer is that most municipalities want development, of housing, industry, offices, etc. So they designate as much land as they think is needed, specifying when they think it will be needed. The result is that building is not seriously restricted by shortages of building land.

This has consequences for prices. Shortages of marketable uses (private housing, offices, etc.) do not arise because of shortage of building land. So the prices of those marketable uses are fairly low, reflecting cost more than scarcity. So the market value of land for marketable uses is fairly low. That is one of the reasons why the question of development gains on land is a "non-issue". In most cases, the volume supplied is so high that development value is not much higher than acquisition costs plus servicing costs (although this has to be placed in the context that the accounting unit is the whole plan area).

The behaviour of Dutch municipalities on the urban land market is expressed very well in the following quotation from a statement of Rotterdam's real estate policy: "As regards the making of profits, a municipal real estate department may in my opinion be compared to a department of public utility, such as for instance the municipal water works, where the primary consideration is to supply a good quality of drinking water at a reasonable price and not to make profits, although some profit at the end of the year would certainly not be unwelcome." (Rotterdam 1959). Although this was written many years ago, it still describes the current attitude (see, for example, similar statements in Giebels et al. (1985: 14) and van den Broek (1988: 85). Particularly revealing is the comparison with a public utility: building land must be as readily available as gas, water, or electricity!

Suppliers of unserviced land
The suppliers of unserviced land are the existing owners: those whose land or land and buildings are acquired by a municipality at the start of a land development process. It is not attractive for these suppliers to develop the land themselves, for technical and financial reasons. The technical reason is that land servicing must take place on a large scale, because most land is unfit for building without drainage, etc. Most private developers could not afford to tackle land development schemes on such a scale. The financial

obstacle is the low gains from development, partly because servicing costs are so high, and partly because so much land is supplied that it has little scarcity value.

Demanders of serviced land

Those who demand serviced land can be divided into three categories. There are non-profit making, semi-public trusts, which build such amenities as social rented housing, subsidized owner-occupied housing, schools, hospitals and clinics, etc. In most plan areas, these take most of the building land (see Ch. 9.2), and they seek to acquire building land easily and at a reasonable price.

Secondly, there are private bodies that build for their own use. These include a few owner-occupiers and most industrial firms. These also wish to acquire building land easily and at reasonable prices.

Finally, there are those who develop property for profit, by building private housing, offices, shops, and very occasionally industrial premises. These wish to acquire building land easily, predictably, and without undue delays at prices that will enable them to make a profit on their developments. The disposal price charged by a municipality might be the full development value of the land, in which case a developer makes a profit only on the construction process. If a municipality charges less for the land, a building developer can also make some development profit on the land. There is no evidence that property developers are unhappy with this situation: they forgo development gains on land (which might in any case be small) but are offered a predictable supply of building land without running any risks.

An idea of the relative importance of the different types of land-use acquisition on greenfield development sites can be derived from figures analyzed in Chapter 5.2.

Land within the plan area not disposed of

As stated in Chapter 4.1, only about half of the land in a plan area is disposed of as building land. The other half is for such amenities as roads, paths, public open space and public car parking. This remains in municipal ownership and is transferred to the relevant department (parks, highways) for management and maintenance. The land is transferred free: it has been paid for out of the income from land disposals.

The rôle of central and provincial governments

Central government can influence the content of a local land-use plan in ways described in Chapter 3.1. It can also subsidize the land development process and until 1990 had to approve the plot prices for Housing Act dwellings. Now it sets guidelines for those prices.

There is a fourth and very important way in which central government can influence the local urban land market. It has a vital interest in the number and mix of dwellings for which claims on a subsidy from central government will be made. This interest is financial, but has strong connections with national housing policy. Therefore there is a system of housing quotas (*contingentering*). Each municipality draws up a housing plan, stating how many houses and of what type it wants to build per year. The dwellings that will receive a central government subsidy fall under the quota system. The province co-ordinates quota requests and sends them to central government. This decides the total number and composition of subsidized dwellings to be built in each province. The provinces then allocate these between their respective municipalities.

Provincial government can influence the content of the local land-use plan in ways described in Chapter 3.1. One aspect must be emphasized here: when deciding whether or not to approve a land-use plan a province must take account of the financial statement relating to the land development process (although there is evidence that they do not do this very strictly). The provinces' involvement in the system of housing quotas is also important, because it is the provinces that allocate the subsidized house-building between the municipalities according to their respective regional plan (*streekplan*). The *streekplan* can also be used to allocate unsubsidized house building between the municipalities.

Another influence that a province can exercise is over municipalities that have fewer than 100,000 residents. These must obtain prior permission from the provincial authorities before buying or selling landed property. When exercising this control, a province looks primarily to the effects on a municipality's finances: the province does not see it as its responsibility to check that the transaction prices are "reasonable".

Property professionals

In each municipality, one or a few housing associations will be active, and these are very important in the local urban land market.. In the larger authorities one or two of these will be very big and very professional. The housing associations commission more housing than other suppliers of housing (see Ch. 9.2) and all the local housing associations in a particular municipality keep close contact with each other and with the municipality.

Housing associations have formed two national associations for exchanging information and representing their interests: the NCIV (Nederlands Christelijk Instituut voor de Volkshuisvesting) and the NWR (Nationale Woningraad).

Property developers are another group of property professionals important for the local urban land market. Sometimes large local builders will develop private housing, but increasingly private property development is in hands of national companies (a few of which work internationally). These have increased their expertise and professionalism in the past few years and have come together in a national association: NEPROM (Vereniging van Neder-landse Projectontwikkeling Maatschappijen). Property development is sometimes undertaken jointly by a property development company and a property investment company: the latter becomes the owner of the property when it is completed, but can also provide building finance.

Municipalities are also crucially important property professionals. Most of the larger ones have strong land departments (*grondbedrijf*) for buying, servicing and selling land. These are often organized as independent agencies. Nationally, there is an association of land departments (Vereniging van Grondbedrijven: VVG). Many of the people working in these municipal departments have received no training in real estate other than that offered by the VVG: the result is that their expertise is narrow.

For further discussion of the rôle of these property professionals, see Chapter 8.2.

4.3 The process of plan implementation

The rôle of the municipality

The double rôle played by municipalities in the urban land market, as both planning authorities and suppliers of building land, has already been described. However, in many cases their involvement does not stop there. Municipalities not only dispose of the land with planning (building) permission supplemented by conditions in the conveyancing contract, they also want be involved in the design and building process. But municipalities actually build very little themselves. How then do they get involved?

The following is common practice in housing areas (which in any case are by far the largest users of building land). A municipality will choose a developer or, for the larger expansion schemes, several developers, sometimes by open tender, sometimes on the basis of "it's your turn next", sometimes on the basis of experience with successful co-operation in the past. The land is transferred to the developer under a contract that stipulates not only the price of the land but also the types of housing to be built (for rental or sale, method of financing, size, etc.), numbers to be built, the price/rent and the delivery date.

Then comes the design and building stage. A smaller authority with a small expansion scheme usually enters into an agreement with one or two developers and lets them get on with the job. A larger authority divides a large expansion scheme into several projects, appointing a developer for each project and forming a building team that contains a representative from the developer, the job architect, an urban designer from the municipality and the building contractor. The team is responsible for the detailed design and for supervision of the realization of the project.

The negotiating culture

It has been said that public affairs in the Netherlands follow a "negotiating culture" or "the politics of accommodation". The latter term is used by Lijphart (1968: 103), who describes accommodation "in the sense of settlement of divisive issues and conflicts where only a minimal consensus exists. Pragmatic solutions are forged for all problems."

This applies equally well to physical planning, both plan-making and implementation. When a plan is being made, a municipality holds intensive consultations with all those involved. These consultations are with the citizens (public participation) and also with other public bodies that must formally approve a plan, other public bodies whose co-operation is necessary for plan implementation, and private and semi-private bodies that could object to the plan or will be involved in its implementation.

The province and the various agencies of central government, plus possibly adjacent municipalities fall into the first category of bodies that need to be consulted. The second category includes the providers of the necessary subsidies and housing quotas (the general directorate for housing), the public utilities, Rijkswaterstaat (responsible for major roads and water management), local water boards, etc. The third category includes housing associations, local chambers of commerce, local associations of shopkeepers, and local environmental groups.

These bodies will often have contradictory interests, which is where another aspect of the negotiating culture becomes apparent. Rather than trying to force through a majority position, a municipality will negotiate until it has discovered a position that is acceptable to as many as possible of the parties involved. This can often take a long time. The duration and complication of all the consultative rounds is a national joke, and something for which Dutch planning is known internationally. But it has a great strength: when the plan is finally completed, most of those involved have committed themselves to implementing it.

During implementation, things can happen that were not expected when commitments were made about plan contents, such as changes in demand and changes in prices. But channels for reopening negotiations are still available, because of the municipality's involvement in design and building, so revised agreements and commitments may be negotiated. In this way, although plan-making takes a long time, plan implementation can be quick.

This negotiating culture as applied to physical planning is manifested, according to ter Heide (1991) by the prevalence of "co-words": co-operation, co-ordination, consultation, consensus, compromise, complexity. Pub-lic–private-partnerships are another manifestation of the negotiating culture that dominates physical planning in the Netherlands.

The frequency of using an instrument

The implementation instruments available to a municipality under public law have been described in Chapter 3.1. The building permit must always be applied for and granted before building can take place, although it is not unknown for the municipality to have made such hard agreements with the developer that the latter is told, with a wink, that he can start to build right away, the permit will come later. (see Thomas et al. 1983: 226 et seq.).

The construction permit is not used much in plans for urban development, although it is widely used in plans for rural areas as a way of preserving the existing situation. Compulsory purchase and the right of first refusal are rarely used.

The implementation instruments available to a municipality under private law, described in Chapter 3.1, are available whenever the municipality itself carries out the land development. The practice is for plan implementation to be by public instruments (mainly, the building permit) supplemented by private instruments. This is what was described earlier as an application of the mixed doctrine of public/private law.

To that can now be added the message of the negotiating culture: plans are implemented not only by the use of formal instruments available under public and private law, but also by persuading many of the parties involved to enter into commitments to implement the plan.

CHAPTER 5

The outcome of
the urban land market

5.1 Changes in the structure of ownership

Effects of the land development process

Land is continuously being taken from rural into urban use and most of this urban land is purchased by municipalities, which dispose of only about half the land they thus acquire. The result is that municipalities own more and more land.

When disposing of the land for building, some municipalities retain legal ownership, disposing of ground leases only. In these municipalities, land ownership grows even faster. A survey in 1981 of all municipalities with more than 25,000 inhabitants, covering more than half of the national population discovered that about 85% of municipalities normally disposed of land freehold. However, 80% of municipalities also used leasehold disposals. And 30% used leasehold frequently, with 12% using it exclusively (de Jonge 1984). Among the municipalities that use leasehold (almost) exclusively are the four big cities of Amsterdam, Rotterdam, Den Haag and Utrecht.

Statistics of land ownership

Statistics of landownership are scanty and give a very partial view. First, it has been calculated (Hollenberg 1978, p. 80) that 1.2 million ha, a third of all land, is owned by public bodies (it must be remembered that this includes most of the land in the IJsselmeerpolders, which is leased to farmers, not sold).

Second, Giebels et al. (1985) have estimated the total area of land owned by municipalities (but excluding the three largest). At the beginning of 1983, this was 100,000ha, or 2.7% of the total land area. Half of this was in plan areas where disposal had not yet started or was not yet completed. However, when municipalities dispose of this land they will also be acquiring more in other plan areas. Their share in landownership will therefore continue to grow.

The share is highest in those municipalities that do not sell freehold land, only ground leases. In Amsterdam, for example, about 80% of land is owned by the municipality (Gemeente Amsterdam 1990). Den Haag is another example, but public ownership of land is particularly high there because of the concentration of central government buildings. The figures in Table 5.1 are taken from Gemeente 's-Gravenhage (1983: 89–90).

Finally, it has been estimated that 44% of agricultural land is owned not by those who work it but by public bodies, foundations, the church, private lessors, etc.(Hollenberg 1978).

Table 5.1 Land ownership in Amsterdam.

	ha	%
area within municipal boundaries	6,795	100
of which: owned by municipality	4,400	65
owned of central government	845	12
owned by others	1,550	23

Source: Gemeente's Gravenhage, 1983, p 89-90

Ownership of developed land

There are no figures for ownership of developed land. The figures presented in Chapter 9.1 give an indication of who owns the building stock, but cannot easily be translated into hectares of developed land (even if we ignore the complication of buildings standing on land disposed of under a ground lease).

5.2 Demand for building land and its supply

Land conversion

Figures for the conversion of land from one use to another were given in Chapter 1.4. An important detail can be added to that information concerning the re-use of industrial land. The extent to which old and disused industrial land is contaminated is becoming clearer and the findings are alarming. So far, 100,000 estates have been found to be contaminated to a greater or lesser extent (ROM, April 1991). These will not be re-used until they have been cleansed, but the costs are enormous and the necessary finance will be made available only slowly. It will come either in the form of subsidies (see Ch. 3.3), costs recouped from polluters, or contributions from land developers. It can be expected that such land will lie vacant for many years before it can contribute to the supply of usable land again.

Transactions on the land market

Sample surveys up to and including 1983 were made of land bought by municipalities for building purposes and of land sold by municipalities and others for building purposes. The results were published regularly in CBS, Maandstatistiek Bouwnijverheid (see also Ch. 3.1). An overview of the results for 1965–83 is presented in Table 5.2.

Table 5.2 Land conversion.

Year	Land bought by municipalities for industrial estates and other buildings (×1,000m²)			Land for industrial estates (×1,000m²) bought from		Land for other buildings (×1,000m²) bought from	
	without building works on	with building works on	total	municip-alities	others	municip-alities	others
1965			5,049	3,071	779	8,645	6,304
1978	7,930	6,550	14,480	5,473	1,577	15,195	5,327
1979	9,288	8,545	17,834	5,377	1,954	12,563	3,177
1980	12,054	7,850	19,904	3,513	2,092	12,011	2,808
1981	9,312	7,107	16,419	1,524	685	11,164	2,567
1982	12,504	3,668	16,172	1,865	1,247	11,086	2,769
1983	8,143	2,585	10,729	1,132	1,020	11,593	4,140

Source: CBS, Maandstatistiek Bouwnijverheid.

Land acquired by municipalities for building purposes (for industrial estates and other uses) is analyzed in Table 5.3, showing type of acquiring municipality. The table shows a decrease in the total area acquired from 1965 (at the time of the large urban expansions) to 1983 (in the economic recession). Also clear is how the weight of acquisition has shifted from rural to urban areas.

Table 5.4 shows for how much land and in how many cases (transactions) compulsory purchase powers were used by municipalities acquiring building land.

The land supplied by municipalities and others for building purposes can also be subjected to a more detailed analysis (see Table 5.5).

Finally, the land supplied by municipalities and others for housing purposes is analysed in Table 5.6 by type of municipality. This again makes clear the declining importance of the rural municipalities.

Scale of supply and demand

The figures analyzed above have not been collected since 1983, but some idea of the current situation on the urban land market is given by Tables 5.7–12. However, it should be borne in mind that it is not possible to give figures for industrial and office uses separately. This is because when building

Table 5.3 Land acquisition by type of municipality.

	A	B	C	D	E	F	total
			Type of municipality				
			area of land acquired (x 1,000 m²)				
1965	13,462	13,405	7,468	8,082	2,918	9,713	55,049
1980 total	6,284	5,576	613	3,270	2,536	1,624	19,903
without building works on	4,127	2,946	241	2,722	1,430	588	12,054
with building works on	2,157	2,630	373	548	1,106	1,035	7,850
1981 total	1,345	6,578	3,443	2,602	1,065	1,386	16,419
without building works on	811	4,437	1,677	1,314	916	158	9,312
with building works on	534	2,141	1,766	1,289	149	1,228	7,107
1982 total	2,611	5,039	3,557	2,311	1,988	666	16,172
without building works on	2,188	3,989	3,221	1,826	893	387	12,504
with building works on	423	1,050	335	485	1,090	279	3,668
1983 total	803	2,671	3,237	1,704	714	1,600	10,729
without building works on	565	1,971	3,126	1,174	3	1,305	8,143
with building works on	239	700	111	529	711	296	2,585

Notes: A rural municipalities.
B urbanized rural municipalities and country towns with fewer than 10,000 inhabitants.
C commuter municipalities.
D small and middle-sized towns (10,000 to 50,000 inhabitants).
E larger towns (50,000 to 100,000 inhabitants).
F large cities (more than 100,000 inhabitants)
Source: CBS, Maandstatistiek Bouwnijverheid

Table 5.4 Compulsory purchase of land.

Year	All purchases		Compulsory purchase	
	no of trans- actions	area ($\times$ 1,000 m^2)	no of trans- actions	area ($\times$ 1,000 m^2)
1965	3,480	55,049	158	3,766
1980	2,488	19,903	–	–
1981	2,572	16,419	–	–
1982	2,760	16,172	48	20
1983	2,412	10,729	8	1

Source: CBS, Maandstatistiek Bouwnijverheid

Table 5.5 Land supplied by municipalities.

	Percentages								
	1965	1976	1977	1978	1979	1980	1981	1982	1983
Building land bought from municipalities									
for industrial use	16	12	17	20	23	17	10	11	6
for other uses	46	54	49	55	55	59	70	65	65
Building land bought from others									
for industrial use	4	6	5	6	9	10	4	7	6
for other uses	34	28	30	19	14	14	16	16	23
				mln m^2					
total area bought		18.8	32.5	32.9	27.6	23.1	20.4	15.9	17.0	17.9

Note: the figures for 1977 do not include information for the cadastral region of Amsterdam.
Source: CBS, Maandstatistiek Bouwnijverheid

Table 5.6 Land supplied for housing.

	Type of municipality						
	A	B	C	D	E	F	total
Bought from municipalities ($\times$1000 m^2)							
1965	2,420	2,793	796	1,103	536	998	8,645
1979	3,706	3,678	584	1,922	1,301	1,372	12,563
1980	3,454	3,093	753	2,485	1,186	1,059	12,011
1981	1,692	3,660	1,866	1,860	1,010	1,075	11,164
1982	1,422	3,283	1,964	2,373	1,091	953	11,086
1983	1,496	3,350	2,491	2,388	745	1,123	11,593
Bought from others ($\times$1000m^2)							
1965	1,686	1,396	334	340	224	234	4,214
1979	1,102	1,283	202	275	172	143	3,177
1980	1,082	895	194	398	94	144	2,808
1981	332	772	281	414	585	183	2,567
1982	381	1,255	306	470	114	244	2,769
1983	1,122	967	815	568	189	480	4,140

Note: the types of municipality are the same as in table 5.3. *Source:* CBS, Maandstatistiek Bouwnijverheid

land is supplied for production purposes it often has a planning designation that allows both uses. Land for production purposes is surveyed annually (see Ch. 3.1) and the results are presented at the level of separate estates. Table 5.7 presents the results at the level of the four regions (for a definition of these regions see Table 1.17).

Table 5.7 Building land supplies.

	1986	1987	1988 ha	1989	1990[1]
North					
area disposed of	93	94	109	146	–
area still available	3,864	3,809	3,718	3,722	3,525
already serviced	1,513	1,546	1,537	1,582	1,525
not yet serviced	2,351	2,263	2,181	2,140	2,000
East					
area disposed of	178	234	267	311	–
area still available	2,529	2,438	2,379	1,939	1,882
already serviced	1,361	1,332	1,211	1,101	947
not yet serviced	1,168	1,106	1,168	838	935
South					
area disposed of	269	425	407	413	–
area still available	5,257	5,262	4,868	4,697	4,388
already serviced	2,782	2,751	2,674	2,570	2,359
not yet serviced	2,475	2,511	2,194	2,127	2,029
West					
area disposed of	260	240	390	519	–
area still available	4,073	4,153	4,121	3,909	3,751
already serviced	2,832	2,823	2,745	2,753	2,610
not yet serviced	1,241	1,330	1,376	1,156	1,141
The Netherlands					
area disposed of	800	993	1,173	1,389	–
area still available	15,723	15,658	15,086	14,267	13,546
already serviced	8,488	8,452	8,167	8,006	7,441
not yet serviced	7,235	7,210	6,919	6,261	6,105

Source: VastGoedMarkt, own recalculations

Note: (1): area disposed of in 1990 not yet known.

Table 5.7 shows the period from 1986. The same source (*VastGoedMarkt*, November 1990) analyzes figures from 1980 to 1990 and draws the following conclusions:

○ In the period 1980–90, for every hectare disposed of another was supplied: in 1980 there were 7,300 ha available, between 1980 and 1990 7,300 ha. were disposed of, in 1990 there were 7,400 ha. immediately available (i.e. serviced);

○ The amount of land available but not yet serviced has decreased from 9,000 ha (1980) to 6,100 ha (1990);

○ The amount disposed of per year fluctuates greatly. The record was in 1977: 1,500 ha. In 1980 it was 770 ha, in 1982–83 around 300 ha (the economic recession) after which it started to grow: it was 800 ha in 1986 and it is now growing at around 200 ha a year;

○ There are big differences between regions, and also within them. In the northern region, about half (1,700 ha) of the area available is in the province of Groningen: the reason is the large area of harbour land, Eemshaven in Delfzijl, for which demand is low. In the eastern region, the area available decreased greatly between 1988 and 1989. This has nothing to do with increased demand: rather, the province of Flevoland decided it had been far too optimistic in supplying industrial land, and redesignated a large area for horticulture. In the western region, the large disposals have been mainly of harbour land: this alone accounts for half of the national increase of 200 ha disposed of each year. In the southern region, disposals have not been growing.

Another source of information for all land uses is the annual survey carried out by TAUW Infra Consult in 79 municipalities. The absolute figures tell us nothing about the national situation, but trends are regarded as being representative for the whole country. The figures are presented with a base of 1982 = 100. The data are collected for greenfield sites only.

For housing, information is presented in Table 5.8 of the number of plots disposed of. This indicates that for single-family dwellings the total number of plots has been fairly stable, although the composition has varied widely, with a strong movement from subsidized to unsubsidized housing. For multi-family dwellings, the number of plots has fallen greatly.

Table 5.9 shows the land disposed of for industrial firms and offices in square metres. It indicates growth.

Table 5.10 shows the distribution of the land disposed of for industry and offices (but also for other non-housing uses) in the 79 municipalities between 1982 and 1986.

Again, the absolute values are of little significance, but the distribution between the various uses is probably representative. Industry takes the lion's share, followed by public buildings and schools.

Oversupply was a characteristic of the 1980s. The Province of Gelderland (in a provincial note, 1983) estimated that at the beginning of 1982 its municipalities had reserves of land for housing sufficient for 10 years, for industry for 10–18 years, and for other uses for 16 years. The municipalities had invested 1.1 billion Hfl in that land and were paying 100 million Hfl a year interest on it.

In the same period, two surveys were made of the supply of industrial land. Ike et al. (1984) discovered reserves in the whole country at the beginning of 1983. The results are shown in Table 5.11.

Olden (1984) carried out a similar survey, the results of which are not directly comparable, but certainly show the same order of oversupply

Table 5.8 The supply of plots for housing.

	Plots for single-family dwellings	Plots for multi-family dwellings
Social sector[1]		
1982	100	100
1983	56	60
1984	59	40
1985	53	17
1986	52	30
1987	58	20
1988	54	22
Market sector with subsidy[2]		
1982	100	100
1983	189	160
1984	337	73
1985	352	94
1986	502	104
1987	252	47
1988	229	23
Unsubsidized[3]		
1982	100	–
1983	315	–
1984	485	100
1985	309	8
1986	405	–
1987	434	41
1988	808	85
Total		
1982	100	100
1983	84	79
1984	111	47
1985	99	25
1986	118	38
1987	99	24
1988	114	6

Source: TAUW 1990

Notes: (1) "HAT", "Woningwetwoningen", "premie-huur non-profit", "premie A"

(2) "premie-huur profit", "premie B", "premie-C huur", "premie-C koop"

(3) "vrije sektor-koop", "vrije sektor-huur"

Table 5.9 Index of land disposed of for industrial firms and offices.

	Industrial firms	Offices
1982	100	100
1983	101	95
1984	105	164
1985	186	173
1986	222	466
1987	291	539
1988	375	105

Source: TAUW 1990.

Note: information from all types of plan area (see the distinction in Table 5.10).

Table 5.10 Land disposed of for industry and offices.

	In plan areas with housing as the main land use	In other plan areas	Total
	%	%	%
schools	18	1	6
public buildings	28	1	8
shops	3	–	1
industrial firms	37	96	81
offices	13	1	4
total	100	100	100

Source: TAUW 1990.

Table 5.11 The supply of industrial land identified by Ike.

	Owned by municipalities	Owned by harbour boards	Owned privately
area (ha)	10,522	2,398	4,015
value (Hfl. mln)	8,622	1,631	not known

Source: Ike *et al*, 1984.

Table 5.12 Industrial land supply identified by Olden.

Land	Served by sea ports	Other
already serviced (ha)	2,960	5,490
not yet serviced (ha)	2,330	6,150
total (ha)	5,290	11,640

Source: Olden *et al.*

(Table 5.12). In 1983, there were reserves of industrial land in the whole of the Netherlands.

There was a large oversupply, especially of industrial land, in the first half of the 1980s. There was even talk of it bringing some municipalities to bankruptcy. This is no longer the case. The economic upturn since then has increased the demand for land, especially industrial land, and the reserves are no longer a financial embarrassment. Clearly supply and demand are in balance again. However, it was a salutary lesson for many municipalities.

Scarcity of building land

The Netherlands is small and densely populated, so it might be expected that building land would be scarce and expensive. The paradox is that it is scarce (or so the politicians say) but not expensive. This is one of the outcomes of the way the urban land market functions and one of the consequences for the urban property market.

Buildings cannot be constructed unless there are plots of land on which they can stand. So supplying land means supplying the possibility for building. If the suppliers of land try to prevent a shortage of building land ever occurring (irrespective of the financial consequences), there will never be a shortage of building possibilities. If, additionally, there is neither a shortage of capital for building nor the will to build, then there will be no shortage of buildings, and average prices for property (and therefore for land) will be low.

Second, supplying a plot of land for building implies supplying not only the possibility of building but also the locational properties that the building will acquire, such as transport infrastructure, quality of the immediate surroundings and proximity to other urban facilities. If the suppliers of land try to ensure that these locational properties are always available (irrespective of cost), then there will be little geographical variation between the plots of land supplied. So there will not only be low average prices (of property and therefore of land), but also little geographical variation in those prices. In the

Netherlands, this effect is reinforced because the country is fairly uniform, and many aspects of government policy and national tradition serve to prevent conditions in one location becoming too different from conditions elsewhere. Some variation is, of course, inevitable: there is, for example, only one international airport and, therefore, one region that offers good access to it. As Chapter 9.3 shows, property prices are high there. On the whole, however, the Netherlands is fairly uniform and the activities of all the municipalities independently supplying building land with good locational properties maintains that uniformity.

So building land is not expensive. But what do the politicians mean when they say that it is scarce? They mean that there is a need for more housing, and a scarcity of sites for it. But if we make the familiar distinction between need and demand, we will understand why that kind of scarcity has not driven up prices: there is scarcity in the sense of unsatisfied need; there is no scarcity in the sense of unsatisfied demand.

5.3 Prices

Agricultural land

The price of agricultural land when it is traded within the agricultural sector is important for the urban land market, because it is the minimum price at which rural land can be acquired for conversion to urban uses (Table 5.13).

Land that is leased out for farming is very much less valuable than land unencumbered by a lease; it can be seen that the annual value of the lease represents only about 1% of the freehold value. The reason is that the legislation (Pachtwet, see Ch. 1.4) gives great protection to the lessees. For this reason, analysis is confined to the prices for full ownership. These also, however, must be interpreted with caution, for they vary greatly according to soil type.

In general, prices for agricultural land are highest in horticultural areas, followed by sandy ground and grazing areas. Land in the fen areas is always cheapest. Sandy ground and grazing areas are relatively expensive because intensive farming (bio-industry) takes place in them. This form of farming grew strongly in the 1980s; it does not need much land but uses it very intensively and requires freehold rather than leasehold tenure.

Another important influence on the price of agricultural land is public policy, including national and EC agriculture and environment policy. The EC policy on milk quotas means that land to which these quotas are attached becomes more valuable. There is a national policy for disposing of slurry from intensive farming, implemented by attaching slurry quotas to land; this also affects the price of land.

Table 5.13 The price of agricultural land.

	1980	1982	1983	1984	1985	1986	1987	1988
Sales price (Hfl per ha)								
including farm buildings[1]								
full ownership	52,000	37,000	40,300	42,000	46,400	52,300	49,800	–
leased	29,700	21,800	23,600	23,000	21,200	27,700	24,900	–
land only, arable								
full ownership	35,700	25,800	29,500	34,200	37,400	39,700	36,600	–
leased	20,000	16,200	17,200	17,800	20,000	21,500	20,200	–
land only, pasture								
full ownership	36,900	26,500	30,400	32,000	37,000	46,300	45,100	–
leased	20,700	15,000	17,300	17,400	18,100	21,100	20,100	–
lease prices (Hfl per ha per year)								
including farm buildings[1]	560	–	650	690	710	735	780	805
land only, arable	440	–	475	515	525	535	560	570
land only, pasture	360	–	380	410	425	430	465	465

Source: CBS, Statistiek overdracht en verpachting van landbouwgronden, in CBS/LEI Landbouwcijfers 1990.

Note: (1) where more than 1 ha of land was available.

The prices at which municipalities acquire building land

Figures on prices paid by municipalities for land which can be transformed into building land were collected and analyzed nationally up to and including 1983 (see Ch. 3.1). These are presented in Table 5.14. The figures for land acquired "without building works on" give the best indication of the price of land itself.

Table 5.14 Land acquired by municipalities for building purposes.

Acquisition price, Hfl per m^2	Without building works on	With building works on	All land
1965	–	–	4.00
1978	8.50	40.70	23.10
1979	10.10	41.80	25.30
1980	10.90	68.50	33.70
1981	10.50	83.10	41.90
1982	8.20	107.60	30.80
1983	9.50	99.10	31.10

Source: CBS, Maandstatistiek Bouwnijverheid

The same figures are analyzed in Table 5.15 according to the type of municipality that acquired the land, and when acquisition was compulsory.

Table 5.15 indicates that prices tend to be a little higher when land is acquired by the larger municipalities.

An overview of the prices paid to municipalities and other bodies when they dispose of building land is given in Table 5.16 . The source is the same as that used in Table 5.15 and the data do not go beyond 1983.

Table 5.17 gives a geographical analysis of disposals for non-industrial uses (including housing). It indicates that disposal prices are lower in rural areas than in urban areas, but the variation is not great, and prices are not always higher the bigger the city.

For disposal prices after 1983 for housing we have to rely on the survey carried out by TAUW Infra Consult (see Ch. 3.1). Although this is based on information collected from only 79 municipalities, it is regarded as representative of the national picture of greenfield development. The basic information collected per housing sector was weighted by the national housing production in that sector. Moreover, for one year (1986) we can compare the results from the TAUW survey with results obtained from a much larger survey (Bovy et al. 1990). This analyzed 366 plans for residential development containing 767 building blocks and 8,482 dwellings. One of the variables

Table 5.15 Land acquisition by type of municipality.

Acquisition price, Hfl per m^2	Type of municipality							
	A	B	C	D	E	F	total	compulsory purchase
1965 total	3.00	3.60	4.10	4.50	5.40	4.90	4.00	3.00
1980 total	17.80	24.30	116.60	33.60	53.00	67.30	33.70	-
without building works on	11.20	8.20	17.50	13.30	6.00	21.20	10.90	-
with building works on	29.20	42.30	180.70	134.60	113.60	93.40	68.50	-
1981 total	22.50	19.00	29.20	30.90	58.80	208.50	41.90	-
without building works on	12.70	8.10	14.40	7.50	11.20	45.90	10.50	-
with building workson	37.50	41.70	43.20	54.90	351.00	229.50	83.10	-
1982 total	15.80	15.40	19.20	32.90	40.10	231.80	30.80	221.70
without building works on	7.20	9.80	6.60	8.00	6.10	17.20	8.20	-
with building works on	60.20	36.80	140.50	126.60	67.80	528.90	107.60	-
1983 total	43.00	20.40	9.50	35.00	55.00	71.60	31.10	649.20
without building works on	17.20	13.00	5.20	10.80	46.60	9.90	9.50	-
with building works on	104.20	41.50	129.10	88.90	55.10	344.20	99.10	-

Source: CBS, Maandstatistiek Bouwnijverheid.
A rural municipalities; B urbanized rural municipalities and country towns with fewer than 10,000 inhabitants;
C commuter municipalities; D small and middle-sized towns (10,000 to 50,000 inhabitants);
E larger towns (50,000 to 100,000 inhabitants); F large cities (more than 100,000 inhabitants)

Table 5.16 The price of building land.

	Land on industrial estates bought from		Land for other building purposes bought from	
	municipalities	others	municipalities	others
1965	13.20	13.70	20.30	12.00
1978	41.60	48.40	90.90	93.20
1979	45.70	48.20	111.20	102.10
1980	52.10	43.90	114.90	121.30
1981	62.20	40.90	125.70	136.70
1982	100.20	40.00	139.50	122.90
1983	90.70	45.30	154.40	125.80

Source: CBS, Maandstatistiek Bouwnijverheid.

Table 5.17 A geographical analysis of prices of disposed land.

			Disposal prices, Hfl per m^2 type of municipality				
	A	B	C	D	E	F	total
acquired from municipalities							
1965	16.30	16.50	29.40	18.50	25.00	33.00	20.30
1979	89.50	108.90	205.60	107.70	102.10	149.60	111.20
1980	99.70	117.40	199.60	101.40	121.00	121.30	114.90
1981	96.20	101.90	161.70	152.10	144.70	126.90	125.70
1982	109.90	116.40	190.20	144.10	161.90	121.40	139.50
1983	112.40	131.50	205.10	155.50	183.40	144.70	154.40
acquired from others							
1965	7.30	11.40	17.80	14.00	15.20	42.60	12.00
1979	89.40	78.00	178.30	125.00	193.30	154.40	102.10
1980	112.10	118.30	187.70	99.60	192.20	133.50	121.30
1981	103.50	114.40	212.90	159.90	106.60	217.70	136.70
1982	110.80	82.80	209.60	135.20	203.30	177.30	122.90
1983	89.50	111.20	144.40	127.60	170.10	188.50	125.80

Source: CBS, Maandstatistiek Bouwnijverheid.

Note: For the different types of municipalities see Table 5.15.

analyzed was plot prices. Its findings on absolute prices, variation between different types of housing, and variation between regions largely agree with the TAUW survey presented below in Table 5.18.

The analysis is for housing land, according to financial category and financial sector (see Ch. 7.2). The qualification should be made that variations in plot prices within one sector can be great.

Table 5.18 shows how prices have changed between 1982 and 1988. For single family dwellings they were fairly steady, in all sectors. For multi-family dwellings there was much more variation. Both characteristics probably spring from public policy for disposal prices for housing land (see Ch. 4.1). For the social sector, prices are kept low and steady. And for multi-family dwellings, the way in which prices per square metre and plot sizes are calculated depends on the number of storeys, which presumably varied through the years.

The steady prices between 1982 and 1988 per plot and per square metre for single family dwellings have a second significance. As Chapter 1.2 indicates, in that same period both GNP per person and disposable income rose steeply (by 34% and 40% respectively, all prices in nominal terms). The conclusion is that land became relatively cheaper for the average person

Table 5.18 Prices of land for housing.

	Single family dwellings			Multi-family dwellings	
	plot size m²	plot price Hfl	price Hfl per m²	plot size¹ m²	price Hfl per m²
Social rented²					
1982–83	145	16,675	115	61	254
1983–84	157	15,857	101	52	251
1984–85	153	16,371	107	49	260
1985–86	150	15,750	105	60	221
1986–87	155	16,275	102	59	212
1987–88	159	15,741	99	45	305
1988	158	15,642	99	60	347
Social sale					
1982–83	168	24,528	146	125	112
1983–84	175	23,800	136	86	246
1984–85	178	23,496	132	89	246
1985–86	189	23,436	124	66	279
1986–87	190	23,750	125	51	364
1987–88	181	23,530	130	43	381
1988	172	23,220	135	46	320
Premie rented profit					
1982–83	156	21,685	139	62	290
1983–84	167	24,716	148	60	299
1984–85	168	21,336	127	58	295
1985–86	169	20,618	122	64	210
1986–87	166	20,916	126	55	277
1987–88	140	19,180	137	25	671
1988	129	17,415	135	27	556
Premie-C sale					
1982–83	210	26,040	124	–	–
1983–84	166	23,240	140	–	–
1984–85	200	26,000	130	148	176
1985–86	205	27,060	132	148	176
1986–87	204	27,132	133	–	–
1987–88	208	27,248	131	271	131
1988	198	26,730	135	195	131
Other free sector					
1982-83	291	38,994	134	–	–
1983-84	272	36,720	135	–	–
1984-85	303	41,208	136	–	–
1985-86	355	48,280	136	–	–
1986-87	377	49,010	130	–	–
1987-88	352	45,760	130	–	–
1988	318	42,612	134	–	–

Source: TAUW, 1990(b).

Notes: (1) own recalculations

(2) this is made up of "HAT", "woningwetwoningen" and "premie-huur-non-profit" - see Ch. 7

Table 5.19 Prices and plot sizes of housing land, 1985–88.

| | Single family dwellings | | | Multi-family dwellings | |
	plot size m^2	plot price Hfl	price Hfl per m^2	plot size[1] m^2	price Hfl per m^2
Social rented					
North	210	13,650	65	63	133
East	159	15,900	100	66	144
South	161	13,486	86	75	173
West	126	17,010	135	49	278
Four big cities[2]	131	17,554	134	37	374
Social sale					
North	251	23,594	94	–	–
East	194	22,310	115	110	114
South	211	22,577	107	–	–
West	155	24,490	158	44	371
Four big cities	141	23,688	168	58	330
Premie rented[3]					
North	161	14,973	93	–	–
East	211	24,265	115	80	143
South	205	22,960	112	67	267
West	134	21,172	158	48	318
Four big cities	114	19,152	168	45	539
Premie sale[4]					
North	283	27,734	98	–	–
East	235	27,260	116	–	–
South	263	27,878	106	148	176
West	167	27,221	163	189	236
Four big cities	111	20,535	185	–	–
Free sector sale					
North	432	40,176	93	–	–
East	361	43,681	121	–	–
South	408	45,288	111	78	221
West	296	53,576	181	67	332
Four big cities	–	–	–	–	–
Free sector rent					
North	289	23,987	83	–	–
East	242	24,442	101	–	–
South	347	40,946	118	–	–
West	133	24,605	185	130	235
Four big cities	–	–	–	–	–

Source: TAUW, 1990(b).

Notes: (1) own recalculations

(2) The "four big cities" are Amsterdam, Rotterdam, The Hague, Utrecht.

(3) Premie rented is "premie-huur-profit" and "premie-C-huur"

(4) Premie sale is "premie-B-koop" and "premie-C-koop"

(he could buy more with his annual available income).

The figures in Table 5.18 show very clearly other results of public policy for housing land: plot prices are lower for housing for rent than for sale; plot prices are lower for housing in the social sector, highest in the unsubsidized sector, in between in the *premie* sector; and prices per square metre are lowest for social-rented housing, but the same for all other types.

Table 5.19 presents average prices and plot sizes in the period 1985-88 according to standard region with the figures for the four big cities presented separately as well.

The price per square metre shows very clear geographical variations. It is lowest in the north, higher in the south, then in the east, highest regionally in the west (around twice the prices found in the north), and even higher in the four big cities. The plot sizes, however, vary more or less inversely. The result is a much smaller variation in prices per plot.

To explain geographical variations we must remember that disposal prices are determined by municipalities, and checked by central government: "the market" exercises its influence only for marketable housing types and only indirectly (through the residual value of land derived from market value of the land plus building). So the geographical differences in prices per square metre are probably caused mainly by differences in acquisition and servicing costs: highest in the west, as more of the land acquired there will be already developed and agricultural land there has to undergo extremely expensive servicing before it can be built upon. Plot prices, however, cannot show much geographical variation. For rented dwellings, central government does not permit it; for *premie*-houses for sale, housing prices do not vary much by region (see Ch. 9.3). The result is, we suggest, that the municipalities as land developers adapt plot size so as to compensate for differences in prices per square metre. The exception is unsubsidized housing for sale, where the market allows higher house (and plot) prices in the west than elsewhere.

The survey by TAUW Infra Consult also analyzed disposal prices after 1983 for non-housing uses, again for greenfield development and again based on information from 79 municipalities. The results are given in Table 5.20.

The high standard deviations mean that only very tentative conclusions can be drawn from Table 5.20. Nevertheless, we suggest the following:

○ Prices for non-housing land in plan areas meant mainly for housing are much higher than in other plan areas. We can suggest no market reason for this. In plan areas for housing municipalities probably charge higher prices as a way of cross-subsidizing prices for housing land;

For the above reason, trends in the prices in housing plan areas do not signify much. If we concentrate on prices in other plan areas, we see a rise over 1982-8 for schools, public buildings and industrial firms.

○ It is notable that the price of building land is considerably less for firms

105

Table 5.20 Land prices for non-housing land.

	Schools		Public buildings		Shops		Industrial firms[1]		Offices	
disposal prices Hfl per m^2	A	B	A	B	A	B	A	B	A	B
1982	117.97	85.00	175.31	37.00	316.30	–	172.70	57.59	146.40	146.77
1983	230.80	74.00	118.55	33.00	366.28	–	144.37	68.99	172.19	100.00
1984	108.83	–	146.42	78.22	251.85	180.00	148.56	56.05	264.96	110.85
1985	175.51	–	145.00	53.89	310.82	–	152.46	63.53	213.75	100.00
1986	167.77	105.85	117.94	55.01	398.58	133.00	69.83	72.03	377.56	–
1987	145.00	–	76.57	205.50	461.50	–	82.79	78.61	154.10	60.00
1988	100.29	103.00	109.85	156.88	583.89	–	74.37	68.92	120.53	500.00
average 1982-88	164.74	85.90	130.78	72.80	358.40	141.02	100.75	69.03	246.05	167.66
average standard deviation[2]	89.72	12.37	74.98	52.94	142.79	17.68	68.11	57.14	152.38	136.59

Source: TAUW Infra Consult, 1990.

Notes: (1) firms in the sectors 2/3, 4, 5, 6, 7 see Ch 1.

(2) the standard deviation can arise because of changes over the years or because of variation within one land use in one year caused by, for example, geographical differences. The bigger the standard deviation (in terms of the average), the less reliable are the conclusions that one can draw from the average figures.

A = land disposed of in plan areas where housing is the main use.

B = land disposed of in other plan areas.

(on industrial estates) than for housing. There are two possible reasons for this. One is that disposal prices are set to cover costs, and that land servicing is cheaper on industrial estates; also the proportion of land retained for common uses is less there. The other is that municipalities subsidize industrial land (or do not charge market prices for it) as a way of attracting firms.

The disposal prices for industrial firms and offices can be subjected to a geographical analysis. The average prices for 1985-8 are used in Table 5.21.

Table 5.21 Land disposal prices for industrial firms and offices.

	Industrial firms		Offices	
	A	B	A	B
disposal prices Hfl per m²				
Region				
north	35.48	25.97	123.92	–
east	84.74	49.92	183.73	100.00
west	130.73	132.08	344.65	500.00
south	58.01	50.02	634.42	60.00
Size of municipality (no. of inhabitants)				
0–20,000	58.21	70.68	363.62	391.80
20,000–50,000	95.73	53.85	143.48	–
50,000–200,000	109.86	85.30	271.23	100.00
more than 200,000	–	–	–	–
Total	80.32	71.42	246.94	364.49
Standard deviation	60.45	61.12	154.96	199.42

Source: TAUW, 1990(b).
Note: see Table 5.20.

Bearing in mind the high standard deviations, our conclusions must be tentative. It would appear that prices are highest in the west and lowest in the north (as they were for housing land), and for industrial firms (but not offices) prices tend to be higher in the bigger municipalities.

5.4 Speculation in land

Pure speculation

By pure speculation is meant the purchase of land in the hope that its value will increase without any further action by the owner. There is no evidence that it is more than an occasional problem. Chapter 2.1 refers to instances of people buying land in plan areas, hoping to gain from the rise in value and to avoid compulsory purchase. This is not regarded as a serious problem. Chapter 2.2 explains how speculation could arise when a municipality disposes of land below its market value to uses that can be sold, and describes how municipalities can counter that by including anti-speculation clauses in the sales contracts.

Risks associated with the land development process

The municipality is in most cases the land developer and, as such, bears all associated financial risks. Until the beginning of the 1980s, most municipalities had not been made aware of this, because the financial results had always been favourable.

The experience of the 1980s, however, has shown how financial risks can arise:

○ costs might be higher than expected;
○ the process might take longer than expected, pushing up interest costs;
○ the market demand might be less than expected;
○ the market price might be less than expected;
○ central government might reduce its allocation of housing quotas from what had been indicated;
○ the municipality itself might change its own urban policy, thus reducing demand for its own land;
○ the competition from other land suppliers (other municipalities) might be stronger than expected.

The experience of the 1980s has taught two lessons that will not quickly be forgotten. First, it showed how difficult it is to forecast demand, mainly because preparations for supplying building land begin so long before land is brought to the market. This process – making the land-use plan, acquiring the land and undertaking large-scale engineering works to service it – takes anything up to 10 years. With such a long production period, demand can turn out differently from what has been forecast.

Secondly, it showed that land development is a huge financial operation with great risks attached. Until the early 1980s, municipalities had usually come favourably out of the land development, and this had lulled many into a sense of financial security. For the first time it became starkly apparent how big the risks are. The land departments of the municipalities are now more aware that land development is more than a routine administrative task: it requires special business expertise and knowledge of land and property markets.

It is not surprising that in a situation where development gains on land are so low, private property developers are happy to leave land development to the municipalities.

CHAPTER 6

Case studies

6.1 Lindenholt, Nijmegen:
greenfield land for housing in a medium-sized town

Context of the development

Lindenholt is a district within Nijmegen, a town in the east of the Netherlands (see general location map) with about 145,000 inhabitants (1991). The initiative for new housing developments comes from the municipality itself, based on its house building programmes. These are formulated in terms of the number of dwellings to be built, specified by financial sector, by building type (semi-detached, terraced, flats, high rise/low rise), and when and where they are to be built. These programmes are based on information collected from the municipality's own housing surveys and from the register of those seeking housing first-time or upgraded housing and the information is interpreted in the light of previous programmes. In Nijmegen, a four-year programme is made and revised every year. Revisions have to take account of changes in central government policy, especially on granting housing subsidies.

Nijmegen has been building actively since the Second World War, and in 1965 crossed a high threshold by starting to build on the other side (the west) of the Maas–Waal canal. In Dukenburg, a new district south of the railway, around 9,500 houses have now been built. It was clear that more new housing would be needed, so plans were also made to build to the north of the railway line – the district of Lindenholt – and also on the other side (the east) of the town (see Fig. 6.1). In 1970 the central government issued a directive (see Ch. 3.1) forbidding the municipality to build in the east because the area was ecologically very valuable, making it necessary to speed up plans for developing Lindenholt.

Those plans were based on the house-building programmes which, as explained, are based partly on the number of people seeking housing. In Nijmegen, this number had been growing rapidly, especially single-person households. In 1970, 6,000 were seeking housing, of which 300 were single-

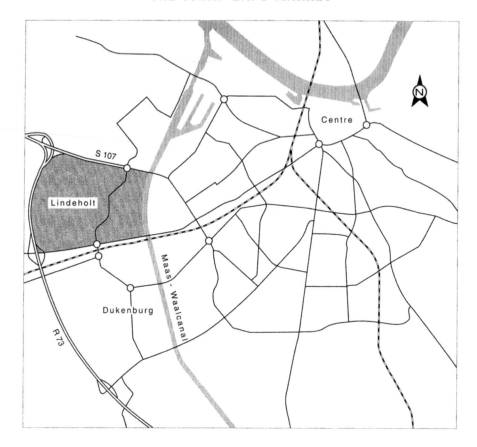

Figure 6.1 The town of Nijmegen.

person households; in 1981, 20,000 were seeking housing, of which 7,000 were single-person households.

When the plans were being made, the housing survey carried out in the city region showed a strong movement out of Nijmegen into surrounding towns and villages. On the other hand, it was central government policy that urbanization should be concentrated. It was therefore decided that Lindenholt should be so designed that it offered a residential environment sufficiently attractive to compete successfully against the outlying villages and small towns. It was also hoped that this policy would encourage richer people to live there, thus keeping their expenditure within Nijmegen. Lindenholt would also be adjacent to the district of Dukenburg; together the two communities would be big enough to support one multi-functional centre for shopping and employment, and colleges for secondary education.

The location for the new district has an area of around 300 ha surrounded by the Maas–Waal canal, the railway line and two motorways, S107 and

R73, which had not been built when planning started, but whose routes had been determined. The location is 8 km west of the town centre.

The land was used for agriculture and was poorly drained, which meant that land servicing costs would be very high. Initially, it was decided that the land most difficult to drain would, for that reason, not be developed. Costs would, however, still be very high, and would therefore attract the lokatiesubsidie from central government (see Ch. 3.3).

The content of the plan

A global land-use plan was published in April 1977 specifying the building density, the ratio of rented/owner-occupied and the ratio of high-rise/low-rise. Of the 300 ha, 209 ha would be used for 5,550 dwellings. The rest would be taken up by office buildings (in the southwest corner), sports fields for local use, and the land retained in agriculture. Half of the dwellings would be for rent, the rest for sale; 20% would be high-rise, 80% low-rise.

This plan was adopted by the municipal council in May 1975. In April 1976, however, the provincial executive announced that it would approve only part of it (neighbourhoods A, B, C, and D). The rest was not approved, as the municipality had ignored the route for the new motorway S107 (this was deliberate as the municipality did not want it built). So the preparation of detailed plans for neighbourhoods A, B, C and D could begin, followed by building. For the rest of the area (Lindenholt-West) the municipality decided (June 1976) to make a new global land-use plan. This envisaged 4,000 dwellings in neighbourhoods of 500 dwellings and a maximum density of 40 dwellings per hectare. This plan would increase the total number of dwellings in Lindenholt to 6,000. For each neighbourhood, a detailed land-use plan would be worked out when it was necessary; in that way, account could be taken of changes in circumstances. In February 1977, the municipality changed the composition of the new housing: it was now to be 40% for rent and 60% for sale, with 10% of the buildings high-rise and 90% low-rise. When the plan itself was published in April 1977 there was yet another change: high-rise dwellings were to make up 5% of construction and low-rise 95%. In April 1978, the plan was adopted by the municipal council.

The plan for Lindenholt-West was approved by the provincial executive in May 1978. It did not acquire legal status until July 1980 because formal objections were still being handled by the Crown, but it had sufficient status for implementation to start.

The different types of dwellings were to be distributed throughout the plan area according to the principle of partial integration, with each dwelling type in small clusters in all of the neighbourhoods. High-rise buildings were to be placed around the main route crossing the district, and expensive housing for sale was sited in peripheral areas free from noise nuisance. House-building was to start in the south-east, next to an existing small centre (St Agneten-weg) and work towards the north-west.

The road system consists of a main route crossing the district south-west–north-east, with neighbourhoods connected to it through two loops. In order to persuade motorists in the west of the district to use the central route, a traffic filter is built into the westerly loop; that hinders private motorists but not buses.

Services were to be provided in three small centres. More would not be necessary, as there is a big district centre for shopping close by on the other side of the railway and colleges for secondary education in Dukenburg just to the south.

Plan implementation

The municipality had considered Lindenholt to be an expansion area long before plan-making started, and as early as 1969 had begun to acquire land there. By the beginning of 1978, before the plan had been formally adopted, more than 90% was in municipal ownership.

We can deduce from the financial statement (see below) that the acquisition costs were on average 7.13 Hfl per square metre excluding demolition costs etc, making 7.45 Hfl per square metre in total. The total costs (acquisition, servicing, etc) were 56.3 Hfl per square metre of all land in the plan area, but there is insufficient information to calculate the total costs per square metre of disposable land. These costs are in line with national averages presented in Chapter 4.1.

The municipality followed the customary land development process described in Chapter 4. Disposal prices were fixed per square metre (unsubsidized) or per plot (subsidized), except for *premie*-C houses (where both methods were used). During the development process, disposal prices have not changed much, and have been held constant since 1983/84. They are:

○ unsubsidized: 125 Hfl per square metre
○ *premie*-C: 125 Hfl per square metre (20,000 Hfl per plot)
○ premie-A: 20,000 Hfl per plot
○ high-rise: maximum of 17,000 Hfl per dwelling
○ social rented: 17,000 Hfl per plot
○ *premie*-B: price per plot 1.25 times the plot price for a *woningwetwoning* (Housing Act dwelling)

These prices are in line with the regional averages presented in Chapter 5.3 (the different categories of subsidized housing are described in Ch. 7.3).

The following qualifications must be added
○ all prices are exclusive of value-added tax;
○ the plot prices for *premie*-A and *premie*-C dwellings are minimum prices.

For plots larger than the standard and with a better location, prices are higher;

○ the total price of the plot on which high-rise dwellings are built is calculated as: for each dwelling on the ground floor 17,000 Hfl, for each dwelling on subsequent floors less than 17,000 Hfl, the price diminishing the higher the floor;

○ building of *premie*-B dwellings has been suspended since January 1988.

The recent (January 1991) state of the finances of the land development process is given in Table 6.1.

The guidelines for development as specified in the global plan for Lindenholt-West were usually modified when the detailed plans were being worked out. This was in response to changing circumstances, such as the temporary collapse of the market for owner-occupied housing, changing housing quotas handed down from central government, and the changing planning policy of the municipality itself. The result was changes in the rate of house-building, and in housing densities and composition. For example, the land that was to have been left open because of the high costs of drainage has now been built upon. Again, it has been possible to build all the *woning-wetwoningen* allowed in the quotas, but not all the *premie*-A and *premie*-C houses for sale, because housing developers were not interested in building them. The modifications can be seen by following the process of house-building.

House building started in 1976/77 in the neighbourhoods A, B, C and D as approved by the provincial executive. The rate of building activity was anything but steady; it began fast, declined with the collapse of the owner-occupied sector at the end of the 1970s, then picked up again. In the middle 1980s it declined again because the municipality was then giving priority to rebuilding the waterfront area in the centre of town and used all its quota of social housing for building there, and developers were not interested in building private housing in an area (Lindenholt) where so much land was lying vacant, awaiting the attention of the municipality.

This pattern can be seen in Tables 6.2 and 6.3. House-building is preceded by the acquisition of building land from the municipality, and in the sales contract the number of house to be built is specified. Table 6.2 shows the number of houses included in the sales contracts signed each year.

Houses are completed a year or so after the building land has been acquired, so the trend shown in Table 6.2 is reflected in housing completions, as shown in Table 6.3.

Table 6.3 shows not only changes in the total number of completions but also in its composition. The stagnation in the owner-occupied sector in the

Table 6.1 The finances of the land development process.

Expenditure	Hfl	Hfl
acquisition costs		21,400,000
attendant costs		439,000
demolition costs		526,000
servicing costs		
drains	18,998,000	
hard surfaces	37,997,000	
public open space	10,971,000	
street lighting and		
fire hydrants	6,596,000	
other works	21,438,000	
		96,000,000
costs of making the plan		14,550,000
contibution to fund for		
'supra-district facilities'[1]		5,982,000
interest payments and cost increases		30,103,000
total		169,000,000
Income		
lokatiesubsidie[2]		6,100,000
land disposals		147,100,000
total		153,200,000
Loss		15,800,000

Source: Municipal records

(1) see Ch 4.1.

(2) see Ch 3.3, This was terminated for Lindenholt in 1982.

Table 6.2 Houses included in sales contracts.

Year	Dwellings	Year	Dwellings
1976	451	1985	158
1977	235	1986	13
1978	117	1987	91
1979	523	1988	163
1980	475	1989	292
1981	1,074	1990	307
1982	841	1991(to May)	99
1983	461		
1984	58	total	5,361

Source: Municipal records

Table 6.3 Housing completions.

	For rent	For sale premie[1]	For sale premie c	For sale unsub'd	Total
before 1978	291	77		14	382
1978	159	191		73	423
1979		188		12	200
1980	329	302		24	655
1981	537	274		25	836
1982	811	101		3	915
1983	396	84	112	2	594
1984	44	70	20		134
1985	15	67	9	1	92
1986	57	48	23	1	129
1987			27	1	28
1988	48	95	45	3	191
1989	153	121	46	19	339
1990	126	56	26	22	230
total	2,966	1,674	308	200	5,148

Source: Municipality records

Note: (1) Before 1981, all premie for sale, after 1981 excluding premie C for sale.

mid-1980s caused financial problems for the land development process. As Chapter 2.1 shows, most municipalities cross-subsidize within a plan area, charging higher than average prices for owner-occupied houses in order to finance lower than average prices for social rented housing. If demand by housing developers for owner-occupied plots is less than expected, total costs have to met by raising more income from plots for social rented housing. In Lindenholt, the municipality did this by a combination of raising the prices of plots and reducing plot sizes (higher housing density) for such housing.

Conclusions

Lindenholt is a good example of the land development process followed in most cases when housing is built on a peripheral greenfield location. It is also a good example of unexpected outcomes. Both plan-making and the plan implementation demanded great flexibility from the municipality. The municipality itself was partly responsible for the changes: it provoked the province by ignoring the route for a new motorway, and halfway through the implementation it gave priority to the waterfront development area. But many of the changing circumstances were beyond its control, in particular changes in the housing market.

Building is now nearly finished, so one can judge the end product. Lindenholt is quite a pleasant residential area, but it is not the enticing living environment which the municipality hoped would attract high-income house buyers. The proportion of rented housing is considerably higher than the 50% specified in 1979, and the municipality has been conducting a publicity campaign to improve Lindenholt's image. Financially, the municipality has made a loss on the land development. It has reserves, however, built up from profits on land development in other plan areas.

6.2 Sloten, Amsterdam: greenfield land for housing in a growing city

Context of the development

Amsterdam is the capital of the Netherlands. It is part of the Randstad, and has a population of about 700,000 within the municipal boundaries (see general location map). In the Fourth National Report on Physical Planning-Extra (1990), the government stated that new housing should be provided as much as possible by redevelopment and by higher densities in existing urban areas. In Amsterdam, it recommended that 20,000 to 25,000 dwellings be built.

For the past few years the municipality of Amsterdam has followed a policy of combined with maintenance and restoration within urban renewal areas, developing infill sites, and building on peripheral locations. It planned to build 7,000 dwellings a year and in the first years of the policy these were to be confined to the existing urban structure. As these possibilities become used up after 1990, attention was switched to Sloten, an area used for market gardening (see Fig. 6.2). This was chosen partly because of its convenient location for many employment areas in the region, its good transport infrastructure,and the use that could be made of existing service centres.

Sloten covers an area of 136ha. Its northern boundary runs through the Slotervaart and along the Plesmanlaan and its southern boundary runs along the Osdorpseweg and the Sloterweg.

The financial situation of the horticultural firms in Sloten was investigated in 1981 (the *Groene Nota*). It was concluded that the area would have to be restructured in the middle term if firms were to remain viable. A designation of part of the area for housing was mentioned as a possibility. A supplementary investigation (the *Gele Nota*, 1982) reported that because of its convenient location, Sloten should be built upon. A feasibility study was made (May 1983) and the possibility (financial and planning) of putting houses there was investigated (November 1984). As a result, it was decided to change the land-use designation.

After a bad experience with the Bijlmermeer development (another large expansion in Amsterdam, which had aimed to provide as many dwellings as

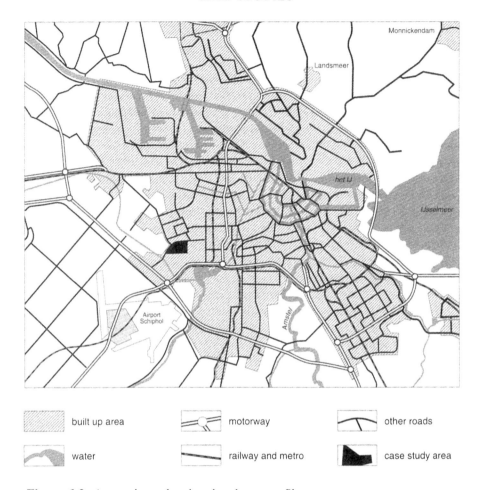

built up area   motorway   other roads

water   railway and metro   case study area

Figure 6.2 Amsterdam, showing the plan area, Sloten.

possible) it was decided that the Sloten development should pay more attention to the nature of housing needs and not just to their size. Recent research into these needs had discovered a pronounced preference for low-rise housing from both families with children and two-person households. Residents desired their environments to be green and on the edge of the city. However, because of the difficulty of finding such conditions in Amsterdam, many people were leaving the city. It was single-person households, young people, and some of the two-person households that preferred to live in the districts near the centre.

The municipality decided to try to meet the preference for low-rise housing, partly to reduce the flight from the city. Dwellings would be built not only for those classified as urgently in need, but also for those with middle to high incomes. This meant much of the development was low-rise single family housing.

The content of the plan

The decision to build housing in Sloten was made formally when the municipal council adopted the structure plan *De Stad Centraal* on 29 May 1985. The land-use plan (*bestemmingsplan*) that followed was a global plan, adopted by the municipality in 1986. Approval was given to part of it by the provincial executive (February 1987), but an appeal was made to the Crown against that approval. On 28 February 1991, the Crown withheld approval from the southern part of the plan because measures had still to be taken to protect that area from noise from the Fokker aircraft factory and Schiphol airport. Apart from these parts, the land-use plan now has legal status.

According to the plan, about 5,500 dwellings are to be built (40 dwellings per hectare gross), 70% are low-rise with gardens, and 30% medium to high-rise. The division between financial categories (see Ch. 7.2) is Housing Act and *premie*-sale A 60%, *premie*-rent and *premie*-sale B 35%, and free-sector sale 5%. What is actually built will depend on the housing quotas allocated by central government to the municipality. For reasons of urban design, the medium- and high-rise dwellings are planned for the centre and along the access routes. The maximum height is 6 storeys, with the exception of the centre, where there will be buildings of up to 10 storeys.

Apart from the housing, other facilities may be built with maximum floor areas of $4,000 \, m^2$ for shops, cafés and businesses; $12,500 \, m^2$ for socio-cultural facilities, social and medical services and sports facilities; and $20,000 \, m^2$ for offices and educational facilities

Plan implementation

By February 1986, the municipality had reached agreement to acquire the land and premises of 97 of the owners and ground lessees, at a cost of 77 million Hfl. In addition, 22 firms had still to be acquired for an estimated 17 million Hfl. After adding other attendant costs, total acquisition costs were therefore, 104 million Hfl.

This gave acquisition costs of 76 Hfl per square metre, which was very much higher than normal. This was because much of the area is covered by glasshouses (negotiations between the horticulturalists and the municipality are not yet finished). The municipality would have been unable to cover the costs of the land development at these prices, so central government was asked to make a contribution. On 21 June 1985 the cabinet announced that it would subsidize the acquisition and servicing costs at 10,000 Hfl per dwelling (using a special provision:*infrastructurele voorzieningen voor nieuwe stedelijke gebieden ter dekking van excessieve verwervingskosten*) with an additional 10,000 Hfl per dwelling as *lokatiesubsidie* (see Ch. 3.3). Part of the agreement lays down that both financial windfalls and setbacks will fall to the municipality, and that it must take all necessary measures to reduce remaining expected losses.

An important change which illustrates the point about changes to subsidies is the decision not to build on another location, Overlanden/de Nieuwe Meer, to the south of the plan area, which was to have had 1200 dwellings. There were many formal objections, so it is to be used instead as public open space. However, one result is that the municipality receives 14 million Hfl less subsidy from central government.

The building programme for Sloten, totalling 5,290 dwellings, is as follows, including the part of the land-use plan yet to be approved:

○ 1990, in quadrants North East and North West 3: 1,700 dwellings;
○ 1991, in quadrant North West 2 and in the centre: 840 dwellings;
○ 1992, in quadrants South East and South West 2: 1,190 dwellings;
○ 1993, in quadrants North West and South West: 860 dwellings;
○ 1994, in quadrants West North West and West South West: 700 dwellings

Two of the quadrants, the North East and the North West, will now be discussed in more financial detail. We need to know the land costs in those parts of the plan area, but it makes no sense to try to allocate actual costs between the parts separately. Instead, the total costs (104 million Hfl) are divided by the total area to be disposed of, and the average costs per square metre are applied equally to all parts. This results in an average cost of 107 Hfl per square metre.

To these costs must be added costs for servicing the land. These include interest, drains, hard surfacing (roads, bike tracks, footpaths, car parking), planting, street lighting and fire hydrants, plan-making and supervision of implementation, and small-scale infrastructure.

The North East quadrant covers 290,700 m^2, so its land costs (290,700 × 107 Hfl) are 31.1 million Hfl. In addition its servicing costs (including value-added tax) are 33.2 million Hfl.

The North West-3 quadrant covers 33,600 m^2, so its acquisition costs (33,600 × 107 Hfl) are 3.6 million Hfl. In addition, its servicing costs (including value-added tax) are 21.1 million Hfl

The combined land costs of the two quadrants are 91.0 million Hfl. In the North East quadrant, 1,300 dwellings are to be built, in the North West, 3,400. So average land costs per dwelling are 53,500 Hfl (including.value-added tax).

When the subsidy of 20,000 Hfl per dwelling is subtracted, the net average land cost is 33,500 Hfl per dwelling. The municipality has estimated the average maximum disposal price of a plot, based on the expected rent or selling price. This is on average 23,800 Hfl. So the municipality will lose

9,700 Hfl on each housing plot.

The municipality intends to cover this loss out of its own reserve funds built up for that purpose. For the two quadrants, 16.5 million Hfl will have to be paid out of that reserve.

Conclusions

Sloten is a striking example of what can happen when land development is the responsibility of the municipality. If we consider the municipality as the producer of building land, then we see that it is producing plots at a cost of 53,500 Hfl each, which can be sold for less than half of that amount (23,800 Hfl). It is justifiable to ask whether this amounts to a responsible production decision? What is the public interest in developing that site, when 56% of the costs are covered by a subsidy?

The question can be asked differently. What would have happened if the land development had been carried out commercially (or, at the least, had not been subsidized). Possibly the market gardeners would have received less compensation, and the building plots would have been sold more dearly. Then Sloten would have been developed for more expensive housing (see Ch. 5.3: the sales prices for plots for free-sector, owner-occupied housing in the west are about 50,000 Hfl). But even that might not have been feasible: horticultural land (especially when it is covered by glasshouses) has a much higher value than agricultural land, so the market gardeners might have refused to sell at a sufficiently low price to make development commercially feasible. In that case Sloten would not have been developed for housing. But the housing need would not have gone away, so other locations with cheaper acquisition and servicing costs would have had to be sought. Probably the location would have been less convenient, with longer (and more expensive) journeys to work.

There is no indication that these alternatives were considered and compared. It would seem that the automatic reaction of a municipality when it acts as land developer is: that losses must be covered by a subsidy (preferably paid by central government).

6.3 Alpha Business Park, Amsterdam: greenfield land for industry

Context of the development

It is the policy of the municipality of Amsterdam and the province of North Holland to reverse the trend of declining employment in the centre of Amsterdam and to strengthen the eminent rôle played by that city. In the

Fourth National Report on Physical Planning-Extra (*Vierde Nota Extra*), the central government supported this policy. Developing the IJ-axis, in particular the central part of it, as a "top location" (see Fig. 6.3), should give the Amsterdam region the best opportunity of becoming the prime location for internationally oriented commercial services. With that, and in combination with other business locations, Amsterdam can offer appropriate conditions to a wide range of businesses.

built up area   motorway   other roads

water   railway and metro   case study area

Figure 6.3 Amsterdam, showing the plan area, Alpha Business Park.

Part of that plan for the IJ-axis is the high quality industrial park known as Alpha Driehoek. This links the banks of the IJ with Teleport, a business area with a telecommunications network. So Alpha Driehoek offers sites that can use the facilities of Teleport but are cheaper than Teleport sites.

Alpa Driehoek also has good road connections and is near a railway station (Sloterdijk). In a few years, connections will be even better, with new roads to Schiphol and to the harbour, and an express tram. The site can clearly been seen from the motorway and the train, and firms can advertise themselves in that way.

The demand for industrial floorspace in the Randstad is strong and has been growing since 1985. The Randstad accounts for about 45% of all industrial floorspace transacted on the open market in the Netherlands. About two thirds of this floorspace is rented, the remainder being sold. The volume of transactions in the Amsterdam region itself has been fairly constant since 1985, accounting for 15% of all transaction volume in the Randstad, and about three quarters of all volume is taken for rent (see Ch. 9.2).

The supply of industrial floorspace offered on the open market for sale or rent has been relatively steady in the Randstad as a whole since 1986, the region accounting for more than 50% of all supply. However, in the agglomeration of Amsterdam, supply has fallen steeply since 1984. In 1989 it was only 40% of the 1984 volume.

The interaction between supply and demand has had its effect on rental levels. These have been estimated by calculating the weighted average of rents for all transactions covering floorspace of 750–10,000 m^2: the results are unreliable as an indicator of absolute rents but probably reliable as an indicator of trends (see Table 6.4).

Rental levels have risen more than inflation (4.9% between 1985 and 1990, see Ch. 1.2).

Table 6.4 Index of rent levels.

	1985	1986	1987	1988	1989	1990
Randstad	100	98	113	115	127	133
Amsterdam region	100	101	107	114	125	131

Source: Rouwenhorst, 1991.

The content of the plan

Alpha Driehoek has an area of 28.4ha, of which 15.7ha is to be disposed of as building land. It lies within the Western Port Area, for which the municipality has drawn up an Action Plan in order to stimulate the economy of Amsterdam. Because of the excellent location of Alpha Driehoek, it has been decided to develop it as quickly as possible to a high quality, and to dispose of it only for high quality activities.

The Alpha Driehoek was originally covered by the land-use plan known as *Overbrakerpolder*. This is a global plan adopted by the municipality on 1

July 1981 and (partly) approved by the province on 12 October 1982. However, in order to be able to reserve the Alpha Driehoek for high-quality activities, a new land-use plan is being made solely for that area, the draft plan of which was published on 27 May 1991.

This new land-use plan imposes strict conditions to achieve a high quality environment. High standards are required of planting, the way plot boundaries are marked, facilities for loading and unloading, outside storage of goods, and the architecture, so that the municipality and the firms can together create the desired quality.

The land-use plan also specifies the types of industrial activity that will be permitted (*staat van inrichting*) in order to exclude firms that might cause environmental nuisance.

Alpha Driehoek is divided into two parts, one for production premises, and for mixed offices/production. In the former, buildings must contain a minimum of 30% office space, rising to a maximum of 70%; in the latter, they must contain at least 70% office space. In the whole estate, no more than 55% of a plot may be built upon, at least 20% must be used for planting, the plot ratio must not be above 1:1, and the maximum building height is set at 19 m. The Alpha Business Park, whose implementation is described in this section, is in the part allocated for production premises.

According to the draft land-use plan, no more than 192,000 m^2 of office and production space may be built in the whole area covered by the plan. If the opportunity to build office floorspace is exploited to the full (100% in the western corner, 70% elsewhere) and the maximum plot ratio is used (1:1), then 140,000 m^2 of offices can be built. What has been built so far does not, however, fully exploit these possibilities.

Plan implementation

All land within the land-use plan for Alpha Driehoek is part of the port area and therefore owned by the municipal ports authority. When it was being serviced to be used for business purposes, no soil contamination was found. In order to make Alpha Driehoek suitable for building it was necessary to raise the ground level, build roads, put in infrastructure (gas, water, electricity), and provide planting and public open space. Public transport is also being improved, and from 1994 there will be a connection with the express tram network.

Site acquisition and development costs have not been made known. No subsidy is being given towards the land development, so all costs will have to be recouped from disposal prices. The land was part of a land-use plan before the new plan for Alpha-Driehoek was made: under the first plan, 35% of the building land had already been disposed of for high quality business uses.

We shall now concentrate on one complex within Alpha Driehoek, the Alpha Business Park. This stands between the Zekeringstraat and the Kabelweg in the south of the site. The developer of this complex is Wilma Vastgoed, which in 1986 acquired the ground lease on 12,111 m², paying a lump sum (premium) of 375 Hfl per square metre instead of a yearly lease over 50 years. Total land costs were therefore just over 4.5 million Hfl.

Wilma Vastgoed started building in 1987 without any units having been let in advance. Construction took place in two phases, to reduce the financial risk: but this proved an unnecessary precaution, as tenants had been found for almost all the space before construction had been completed.

The complex contains five pavilions: two with one storey, three with two. Total lettable area is 10,620 m², of which 4,210 m² (40%) is for production and 6,410 m² (60%) for office use. The space for production is only suitable for light production: ceiling height is 3.6 m instead of the usual 6.5 m; and floors can carry a maximum weight of 1000 kg per square metre instead of the usual 2200 kg. The space can be used very flexibly: partitions can be installed and ceilings in the production space lowered with panels to convert it to office space. (The maximum permissible office space of 70% applies to the whole complex, not to individual pavilions.) Almost all the tenants have made use of the conversion option. In this way, the 70% maximum on office space applicable to the whole complex has in fact been exceeded.

Much attention has been paid to the quality of the architecture and to giving each building a separate identity. By including so much planting, it is hoped that the value of the complex as a whole will be raised. On the site there are also about 170 parking spaces.

Construction costs were 1,100 Hfl per square metre, a total of 11.5 million Hfl, excluding professional fees etc. These costs per square metre can be compared with the national average of 350 Hfl to 400 Hfl (see Ch. 9.2). Tenants were sought in the computer hardware and software, electrical engineering, and distribution sectors. The tenants were selected not only for their financial status; it was also decided they should not be competitors of other tenants. Moreover, the high rents should exclude unwanted tenants. Two of the five pavilions have been let to the PTT (the Dutch telecommunications utility).

In 1989, annual rents were 175 Hfl per square metre of floorspace for production space and 195 Hfl per square metre for office space. This distinction has now been dropped: all rents are 195 Hfl. This rental can be compared with average office rents of around 250 Hfl per square metre in the region. Most of the tenants are using the premises primarily as offices and are therefore paying below the local rate for offices.

The developer sold the complex in June 1989. The gross annual income at the time of the sale was 1,987 million Hfl and the developer asked 26 million Hfl (an initial gross yield of 7.6%). Some Dutch investors were interested, but only at a price that would have given an initial yield of 8% to 8.5%. A Finnish investor was prepared to accept 7.8% and paid 25.5 million Hfl. The developer had spent 16 million Hfl on land and construction costs, and was happy with the sale. He is now developing a similar complex (Omega Park) on an adjacent site, which will have an even higher proportion of office space.

Conclusions

The Alpha Business Park has been a successful venture for the property developer. It is too soon to say whether Alpha Driehoek as a whole will be successful for its developer, the municipality of Amsterdam. Will the goals set in the land-use plan be realized? Will the municipality cover its land development costs? Considering the financial success of the property developer, could the municipality not have charged higher disposal prices? The answers to all those questions will be positive if all projects are as successful as the Alpha Business Park.

PART III
The urban
property market

CHAPTER 7

The framework within which the urban property market functions

7.1 The legal environment

Legislation
Some of the legislation relating to urban property has already been described in the section on the urban market relating to new building (see, for example, Ch. 3.1). Of particular importance is the system of building permits. There is also a substantial body of legislation relating to the use of existing property, to be described here. Most of this relates specifically to housing.

Legislation regulating the use of housing   Under the Dwelling Space Act (Woonruimtewet, 1947) it can be illegal to occupy a dwelling or to offer a dwelling for occupation without a permit (*woonruimtevergunning*) issued by the municipality. This law was passed when housing was very scarce, with the aim of fair allocation. Since then it is been applicable only in certain municipalities where the greatest scarcity occurs and, in these, only for dwellings with a rent or sales price below a specified level. This is because those people who need protection through the allocation process would not occupy expensive housing. A municipality can use the following criteria for allocating housing: urgency, relationship between household size and income and dwelling size and price, and social and work connections between the household and the municipality.

Under the same legislation, the municipalities can requisition dwellings and other buildings for residential use. The Vacant Buildings Act (Leegstandwet, 1985) gives further powers to bring vacant buildings into use. Requisitioning is rarely used, and the Vacant Buildings Act is used mainly against squatters.

Under the Housing Act (Woningwet, substantially revised in 1984) a municipal council can declare a building unfit for human habitation (*onbewoonbaarverklaring*): it must then be vacated within six months.

Under the same legislation, it is possible to prevent a building being taken out of residential use without permission (*woonruimteonttrekking*). This power can be applied only in municipalities indicated for this purpose by the Crown. The decision to grant permission is for the municipality to take; it can grant it on condition that the applicant pays into a compensation fund or provides alternative dwelling space.

The Housing Act allows a municipality to require the owner of a dwelling unsuitable for residential purposes to take measures to bring it up to required standard, by means of a directive or building order (*aanschrijving*). The occupant must allow the improvements to take place, and they must be completed within a specified period. If the owner does not comply, the municipality can carry out the works at the owner's expense.

Under the Housing Act, supplemented by specific legislation (Wet op de woningsplitsing, 1987), it is in certain circumstances forbidden to divide an existing building into apartments. This applies only in certain municipalities or parts thereof, and only for buildings built before a certain date, both conditions to be specified by the Crown. Implementation is for the municipality.

The Housing Act also allows a specified proportion of those rented dwellings built with the support of a central government subsidy (see Ch. 7.3) by municipalities or recognized housing associations to be reserved for special groups. Certain classes of public servants and railway employees who are obliged to remove to a municipality can be allocated 5% of such housing, and another 5% can be reserved for refugees and those requesting asylum.

The sale of social rented dwellings (Housing Act and *premie*-rented dwellings, see Ch. 7.3) is regulated under a decree concerning recognized housing associations (Besluit toegelaten instellingen volkshuisvesting, 1976). In order to qualify for purchase, a household must have rented for at least five years and have an income below a specified level. Minimum sales price is the market value of the dwelling. There is no right to buy, nor an obligation to sell. The municipal council must approve all sales.

A more detailed account of housing legislation is to be found in Min. VROM (1989d).

Landlord–tenant legislation for housing Legislation on landlords and tenants is in two parts. The first regulates their relationship. The general rules relating to all landlord–tenant agreements are to be found in the civil code, Book 7A, Title 7; the rules specific to housing in the Articles 1623(a)–(o). The rules are binding: contracting parties are not entitled to deviate from them. The rules cover the way a contract may be terminated, how it should be continued if the tenant and his or her spouse divorce, what happens on the death of a tenant; and what happens if two tenants want to exchange tenancies. Termination of a contract must be announced by registered letter, the length of agreed notice must be observed, and grounds for termination must be specified. If the tenant objects, termination of the tenancy and his vacating the dwelling become matters for court decision.

The second part of the legislation on landlords and tenants regulates the level of housing rents and is contained in the Housing Rents Act (Huur-prijzenwet Woonruimte, 1979, revised in 1988). For rented dwellings that are subsidized by central government, the initial rent is set by the minister (see Ch. 7.3). For the subsequent five years, there is an obligatory annual rent adjustment (increase) with the percentage increase set by the minister. In all other cases, landlord and tenant agree on a rent, and this can be revised annually. If landlord and tenant cannot reach agreement, they can ask the rent committee (*huurcommissie*) to give an opinion. Thereafter, both landlord and tenant have two months within which they can ask the court (*kantonrechter*) to determine the rent.

For dwellings occupied for the first time since 1 July 1989 and with a rent of more than 750 Hfl a month, the rent committee and court must try to set the rent at market value. In all other cases, they must take account of "prescriptions" specified in the relevant decree. These prescribe ways of determining rents (based on the facilities offered by the dwelling), annual rent adjustments, etc.

When rented dwellings have been improved with the help of government subsidies, there are rules about by how much the rent may be increased after improvement (see, below).

Landlord–tenant legislation for non-residential properties For the purpose of landlord–tenant legislation, non-residential property is divided into two types: "1624 business premises" and other business premises. The first type is specified in Article 1624, Book 7A of the civil code: it consists of business premises that serve customers who have to visit the establishment itself (the goods and services cannot be delivered to the customer). It includes shops, cafés, tradesmen's premises, etc. The other type of business premises includes factories, warehouses, offices, banks, etc; premises for use by professionals such as dentists, lawyers and doctors; and all other premises other than residential, such as club rooms and sports halls.

The legislation for premises that are not residential and do not come under Article 1624 is to be found in Book 7A of the civil code, Articles 1584–1623, which regulate all landlord–tenant relationships, and in the Rent Act (Huurwet, revised in 1981). This Rent Act does not apply to residential premises, "1624-premises" or undeveloped land. The legislation is discussed in Smit (1989).

The relationship between landlords and tenants of non-residential and non-1624 premises is determined primarily by the contract to which both voluntarily bind themselves, and whose content is primarily a matter for the two parties alone. That is the principle laid down in the Civil Code.

The other legislation regulating this relationship is in the Rent Act. According to this, the level of rent is purely a matter for the contractual partners. Common practice is to agree to an inflation-linked rent plus rent-revision after five years if the rent has grown out of line with market levels. The contract usually also incorporates provision for paying service costs.

The expiration of the lease is in itself not sufficient to oblige a tenant to quit the property: the landlord must request that separately and explicitly. The premises must be vacated within two months after notice to quit (i.e. there is two months' grace). If the tenant does not apply for an extension during the grace period, then he has to quit when it ends. If the tenant does apply for an extension, this can be granted. But it is not an extension of the tenancy agreement that is granted but an extension of the obligation to vacate. Such an application can be refused because it has not been made in time, because the interests of the landlord would be damaged more by an extension than the interests of the tenant would be by having to vacate, or because of non-observance of the tenancy agreement during the period of the tenancy.

Appeal against the court's decision is possible only on a point of law. Because it is not the lease itself that is extended, if the obligation to quit is postponed, then the court must fix compensation (in lieu of rent) payable by the tenant to the landlord. If the application is granted, it is for a maximum of 12 months (in practice, 12 months minus the 2 months' grace period). If the tenant repeats the application, it can be granted at most two more times, each for a maximum of one year. If the tenant himself wants to terminate the agreement before the contract period has ended, there is no two months' grace period. If the premises are in a "non-liberalized area" and if the monthly rent is 100 Hfl or less, then other rules apply. The number of such premises is small.

The "1624-premises" are treated differently mainly because their business is dependent upon the location of the premises, so continuity in this is of vital importance. Because of the strong link between business operators and their premises, special legislation has been passed to safeguard tenants' positions. Moreover, these safeguards are specified in the civil code in such a way that contracting parties are not entitled to agree to deviate from them. However, in spite of this and although the Rent Act (Huurwet) does not apply, the principle that the price is not regulated does: the rent is a matter for the contracting parties.

There are several important safeguards of a tenant's interests:
○ the minimum tenancy period is 10 years, split into 5 + 5 (6 + 4, 7 + 3, etc are also possible). After the first 5 years, the tenant has an unlimited right to terminate the agreement and the landlord a limited right. If the

tenant wants to terminate after the first period, he must give a year's notice (i.e. hand it in after 4 years). If he does not do this, he is bound to the full 10 years. The landlord can be granted a request to terminate after the first period on only two grounds: he wants to use the premises himself personally and permanently as 1624-business premises and his need is urgent; or the tenant has not fulfilled the terms of the agreement. The landlord must make this request before the end of the 4th year (ie a year's notice must be given). If the ground is that the landlord himself wants to occupy the premises, then 3 years' notice must be given. If the application is granted on this ground, the landlord must pay the tenant compensation. If the landlord's request is not granted, then the tenancy runs its full 10 years;

○ at the end of the 10-year tenancy period, the tenant can request an extension. This can be granted if the landlord has no great financial interest in terminating the agreement, and if the tenant would suffer loss by having to vacate;

○ at the end of the 10-year tenancy period, the agreement does not terminate automatically: explicit notice must be given a year in advance;

○ if the tenancy agreement is for two years or less, then the above rules do not apply;

○ if the landlord sells the property, this does not break the lease contract: the new landlord must abide by all the rights and obligations of the seller. (In this way, although a tenancy to rent is a right *in persona*, it has been given some of the properties of a right *in rem*, see Ch. 1.1);

○ if landlord and tenant cannot reach agreement about the level of the rent, either can ask the court (*kantonrechter*) to determine it. This is not possible at the time the contract is first being entered into, but it is possible when the contract is to be extended and also once every five years if the contract is for an indefinite period. In these cases, the court may act only if the rent is "out of line with comparable premises in the same area". All the regional chambers of commerce have set up advisory committees (*bedrijfshuuradviescommissies*), which courts can consult (but without obligation to do so) when determining rents.

Other legislation relating to the use of buildings The demolition of buildings can be regulated in various ways. In those areas where the taking of a building out of residential use is regulated (*woonruimteonttrekking*, see above), demolition would have this effect and so falls under the regulation. If the building is a registered monument (see below) demolition without a permit is prohibited. And even if a historic building is not so registered, if the province or municipality have regulated accordingly in their monument ordinances (see below), then permission to demolish is required. If the building is in a conservation area (*beschermd stads- en dorpsgezicht*), permission to demolish can be required if specified in the land-use plan for the area. Finally, the Housing Act requires that each municipality apply building regulations (*bouwverordening*, see Ch. 3.1), which also contain conditions regulating demolition. At the least, intention to demolish must be announced to the municipality and demolition may not begin until receipt of that announcement has been acknowledged. But it may be further specified that demolition is not possible without a permit (Adriaansens & Fortgens 1986).

Buildings can be registered as monuments if they are over 50 years old, on the basis of several criteria, as a result of which demolition is regulated (see above) and they can receive subsidies (see Ch. 7.3). The listing of individual monuments that are to be legally protected is a matter for the ministry of welfare, public health and culture, through its agency (Rijksdienst voor de Monumentenzorg). The designation of town and village conservation areas is a matter for that minister together with the ministry of housing, physical planning, and environment. Municipalities can decide about changes to the listing of monuments, and even about demolitions (although the minister of culture may overrule a demolition decision): but to exercise such powers, a municipality must have introduced a set of local ordinances and established a monuments advisory committee. If it does not do this, powers remain with central government.

Environmental protection as it affects new building was described in Chapter 3.1. It should be noted that much of that legislation refers to the processes in those premises; changes in processes in existing buildings fall under the legislation too.

The ways in which a governmental body may intervene in the land market using public and private law are described in Chapter 3.1. In respect of the property market, it is necessary to add that a government body can intervene using private law in those cases:

○ where that body is owner of the property. Sometimes this is the only effective way of achieving a policy, an example being the improvement of privately rented dwellings, if offering subsidies or issuing a building order are insufficient to induce the landlord to carry out the necessary improvements. Then, the only option is for the public body itself to acquire the dwelling;

○ where the governmental body is the landowner and has disposed of ground leases on it. The possibilities that this gives are described in Chapter 3.1. However, the public body can be challenged under the principle of responsible administration (*behoorlijk bestuur*) if it tries to use its power as ground lessor to regulate in contravention of existing public policies (e.g. to forbid something permitted in the land-use plan).

Finally, it should be added that private bodies that have mutually agreed a contract under private law may call in the judiciary to intervene if one or both of the parties think that the contract is being violated or misused, as outlined above.

Information systems

Information systems relating to land are described in Chapter 3.1. Most of the information relevant to developed property is included in the same systems, so only a few points need to be added.

The last time a population census was held was 1971. Basic demographic statistics can still be compiled using the municipalities' registers of their inhabitants, as it is compulsory to register in the municipality of residence. Information about housing conditions, often collected by a population census elsewhere, is now collected separately from a small sample survey, the *Kwalitatief Woningonderzoek*.

The private journal *VastGoedMarkt* sometimes publishes reviews of the various sectors of the property market, such as offices, shops and industrial premises. Most estate agents are members of a national association (Nederlandse Vereniging van Makelaars in Onroerende Goederen, NVM). Members are obliged to submit statements of all properties on their books and of all transactions to the association, which analyzes the data and publishes the results.

House prices as registered with the cadastral service were until 1985 analyzed by the CBS and published in the journal *Maandstatistiek Bouwnijverheid*. Average prices from this source are lower than from the NVM. This is because the CBS analyzes the sales of all houses, whereas the NVM analyzes only prices of houses sold by its members; the cheaper houses are often sold without the aid of an estate agent. Since 1985, house prices are no longer analyzed by the CBS but by the cadastral service and published as KADOR statistics. But this analysis excludes individual sales of less than 40,000 Hfl and more than 400,000 Hfl.

7.2 The financial environment

Financing of property development, ownership, and use is as described in Chapter 3.2. Here, additional material on the housing sector is given. Dwellings are grouped into three "financial categories", according to the way in which building is financed. These are Housing Act dwellings, *premie* dwellings and free sector dwellings.

Housing Act dwellings (*woningwetwoningen*) were financed with loans provided by central government at market interest rates, as regulated under the Housing Act, hence the name. Since 1988, however, finance has had to be obtained by commercial loans, but with a guarantee provided by the municipality in which they are to be built. The guarantee reduces the risk and hence the interest rate. Such dwellings are commissioned by municipalities, by housing associations and (rarely) by private investors, pension funds, etc. They are built for rental, although they may be sold later, and must satisfy specified technical standards and cost requirements.

Premie dwellings are financed with commercial loans or out of reserves and are guaranteed by the municipality. They are commissioned mostly by housing associations, but also by municipalities and private investors. They are built both for sale and for rent, and must satisfy the same technical standards as Housing Act dwellings. Until 1988, there was a difference between *premie*-rented and Housing Act dwellings in the way they were financed. With the suspension of government financing of Housing Act dwellings, however, that difference no longer exists.

Free-sector dwellings are financed with commercial loans, risk-bearing capital, or out of reserves. There is no government involvement. Most are built for sale. They are commissioned by private developers (22%), investors (6%), building contractors (36%), or private individuals who mostly occupy the buildings themselves (32%). The figures in brackets are an estimate of the proportion of all free-sector dwellings built in 1986 (De Rooy & Elbers 1989, p. 7). In principle, free-sector dwellings are eligible for the subsidies granted to Housing Act and *premie* dwellings, but usually cost more than the limit above which subsidies cease.

Transaction costs are calculated in the same way for developed property as for land, as is described in Chapter 3.2.

7.3 The tax and subsidy environment

Taxes concerning the property market

Most of the taxes levied on developed land are the same as those levied on undeveloped land, which have been described in Chapter 3. There is only one tax that is specific to developed land, and then only to housing.

The owner of a dwelling, if he is also the occupier, is regarded as enjoying an imputed rent. This is added to personal income and is therefore subject to income tax (*huurwaardeforfait*). The imputed annual rent is fixed at around 1% of the value of the dwelling (under 1% for cheaper dwellings, more for more expensive dwellings). The government intends to raise this percentage.

Subsidies concerning the property market

Tax deductibility A number of outgoings connected with real estate are tax-deductible, and can therefore be regarded as subsidies. They are interest payments and depreciation payments.

The deduction interest payments applies on all loans, against both personal and company income tax, although there can be small differences depending how a company is registered: Naamloze Vennootschap (nv), Besloten Vennootschap (bv), etc. All costs necessary for acquiring finance (interest payments, lawyers' fees, etc.) are tax-deductible, but not the costs necessary for acquiring the property (conveyancing costs, transfer tax, valuation fees, etc).

Depreciation payments are deductible for company income tax, but not personal income tax. Tax is levied on profits, so depreciation can be set off against them. The tax rules regulating the deductibility of depreciation payments provide that the historical cost is the basis of depreciation; that depreciation is over the economic life of the building, which varies with the nature and quality, but will not exceed 40 years (Andersen 1991, p. 45); and that depreciation is not deductible on the residual value of the property and therefore not on the value of the land.

Housing subsidies The distinction must first be made between object and subject subsidies. The former are given in order to reduce the costs of the dwelling, whoever occupies it, and the latter direct to the occupier.

OBJECT SUBSIDIES FOR RENTED DWELLINGS The first principle is that each complex built should be self-financing, with or without subsidy. Cross-subsidizing between complexes is not permitted. The second principle is that the subsidy is given to reduce operating costs, which consist of capital charges on the loan (borrowed for 50 years), payment into a reserve fund, and maintenance and management. Capital charges are determined by total development costs and the interest rate, the reserve payment is set at 0.5% of total development costs, and maintenance plus management at 1.2%. Because a rent that would cover total operating costs would be too high for many tenants, these costs can be reduced by a subsidy.

Previously, a lump sum subsidy was given, but that took no account of inflation (tenants should be able to afford increasingly high rents over the life of a dwelling). In 1975 a new subsidy system was introduced (Beschikking Geldelijke Steun Huurwoningen). First, the level of rents necessary to cover operating costs without subsidy has to be calculated by a "dynamic method". Assuming a certain level of inflation, the rents can be raised over the life of the dwelling (50 years) in such a way that operating losses in the first years, and the interest on those losses, are covered by profits in later years. The initial rent calculated in this way is lower than that calculated by a non-dynamic method (and this was the aim: the government wanted to reduce housing subsidies). A fair rent is calculated as a percentage of total development costs. That percentage is dependent on the number of rooms but also on total development costs. Above a certain maximum cost, no subsidy is given: it is this which excludes most free-sector rented dwellings from a subsidy. If the initial rent is higher than the fair rent, the difference is covered by an annual subsidy, which is reviewed every 10 years.

There are no national rules limiting subsidized rented dwellings to people with incomes below a certain level. However, every municipality is obliged to make and apply ordinances about how it will allocate dwellings to applicants, and these will include reference to the income of the applicant.

OBJECT SUBSIDIES FOR OWNER-OCCUPIED DWELLINGS *Premie* dwellings built for sale can receive a subsidy under the Beschikking Geldelijke Steun Eigen Woning 1984. *Premie*-A are the cheaper dwellings (e.g. a maximum of 142,000 Hfl in Lindenholt in 1987, see Ch. 6.1) and subsidy is dependent on the taxable income of the buyer (e.g. in 1987, maximum 51,000 Hfl pa): the subsidy is constant (e.g. in 1987, 4,000 Hfl per annum) for a maximum of 30 years. If the dwelling is sold within 10 years, a part of any price increase must be paid to the municipality. *Premie*-B are more expensive dwellings, and the subsidy is dependent on the price: it is paid yearly for between two and five years. The *Premie*-B subsidy was withdrawn at the beginning of 1988.

A subsidy can also be given to the purchaser of free-sector dwellings built for sale (Beschikking Geldelijke Steun Bijdrage ineens Vrije-sector woningen, 1983). These are called *Premie*-C dwellings (although they are not *premie* dwellings in the sense defined in Ch. 7.2). If the price is below 190,000 Hfl and the taxable income of the buyer below 71,000 Hfl pa, a lump-sum subsidy (e.g. 5,000 Hfl in 1988) is paid to reduce the purchase price (Min. VROM 1989d, p. 37).

OBJECT SUBSIDIES FOR ALL DWELLINGS A subsidy (*Regeling bijdrage ineens bouwplaatskopkosten*) was introduced at the beginning of 1989 that aims to stimulate new housing construction within built-up areas. In such areas, construction costs can be higher than elsewhere. A lump-sum subsidy is available so that location-specific costs, which do not contribute to a higher housing quality, do not result in higher rents.

Subsidies for the land development process, described in Chapter 3.3, reduce the costs of all dwellings built in a relevant location.

SUBJECT SUBSIDIES FOR RENTED DWELLINGS   Under the Act Individuele Huursubsidie 1986, all tenants are in principle eligible for a rent subsidy, irrespective of the "financial category" of the dwelling and of whether that dwelling receives an object subsidy. A "norm rent quota" is specified, being the percentage of a tenant's taxable income that he can be expected to pay on rent; the lower the income, the lower the quota. A tenant in a dwelling with a rent higher than the level he can be expected to pay according to his norm rent quota can apply for a subsidy, but the whole of the difference is not covered by the subsidy (otherwise there would be no incentive for people with lower incomes to take cheaper properties). The subsidy depends also on the composition of the household. Above a certain annual income (e.g. in 1988, 36,000 Hfl) the subsidy is not paid.

Two other subject subsidies can be mentioned, although they are much less common. In order to encourage tenants who can afford higher rents to move to more expensive dwellings (thus freeing cheaper dwellings for poorer people, and reducing the claims for subject subsidies), part of the higher rent can be subsidized for a few years (a habituation contribution: *gewennings-bijdrage*). And tenants who must rehouse because of housing improvements or slum clearance can claim a grant towards removals and furnishing costs.

Tax relief on mortgage interest   It has already been stated that interest payments are tax deductible. This also applies to interest payments on mortgages, and there is no limit on the amount of interest that is tax-deductible. This can be regarded as a subject subsidy: it is very regressive, being higher the higher the income of the lender and the higher the mortgage. The example in Table 7.1 makes this clear.

Table 7.1 Tax relief and mortgage finance.

	Hfl	Hfl
taxable income without interest relief	30,000	50,000
tax due	7,512	16,160
mortgage interest (10% on 100,000 Hfl)	10,000	10,000
taxable income becomes	20,000	40,000
tax due becomes	4,068	11,229
tax reduction	3,444	4,931
mortgage interest (10% on 200,000 Hfl)	20,000	20,000
taxable income becomes	10,000	30,000
tax due becomes	1,434	7,268
tax reduction	6,078	8,892

Note : tax system as it was in 1988

The regressive effect of this tax measure should be partly redressed by the tax on imputed income from owner occupation, but the level at which that tax is set is so low that it has a negligible effect.

Subsidies for housing improvements Subsidies towards the improvement of owner-occupied housing used to be paid for each case separately by central government, but since the Town and Village Renewal Act (see Ch. 2.1), this has changed. Central government finances an urban renewal fund which is distributed between provinces and municipalities. These authorities can decide how to use it to subsidize such improvements. Each municipality has to make its own ordinance for this. The municipality of Nijmegen serves as an example: the dwelling must be a permanent one, older than 25 years, and the improvement must cost between 1,500 Hfl and 90,000 Hfl. The subsidy is determined for each case separately, and is paid as a lump sum.

Subsidies towards the improvement and structural maintenance of rented dwellings are paid by central government under the *Regeling geldelijke steun voorzieningen aan huurwoningen*, 1987 or, in urban renewal areas, out of the urban renewal fund. In both cases, central government sets the guidelines, but municipalities (or the provinces in the case of the smaller municipalities) decide after consultation with central government. A lump-sum subsidy is available, varying from 0% to 80% of improvement costs. The percentage depends on the costs of the improvement as a proportion of the costs of new construction, and also on the age of the dwelling. Part of the unsubsidized costs may be recouped by raising rents. From the beginning of 1992 this subsidy scheme will be replaced by another (Besluit Woninggebonden subsidies). Subsidies will then only be available when the costs of the improvements are more than 50% of the replacement value of the dwelling being improved.

The cost of housing subsidies The following estimates of the cost of housing subsidies to the public exchequer in 1991 are given in VROM 1990: all object subsidies for rented dwellings, 5.007 billion Hfl mln; all subject subsidies for rented dwellings, 806.3 million Hfl; subject subsidies for tenants of rented dwellings, 1.85 billion Hfl.

Much less political attention is paid to the cost of giving tax relief on mortgage interest payments (tax revenue forgone). Some sharp questioning in parliament produced the information that tax relief on mortgage interest (estimate 1990) cost 6.5205 billion Hfl (Woonbond 1991). It can be seen that the cost of subsidies for rented housing (55% of all dwellings) is less than the cost of subsidies for owner-occupied housing (45%).

The recipients of the subject subsidies can be compared. In the year 1988–9, 919,000 subsidies were granted to tenants, with an annual average value of 1,755 Hfl. By contrast, 2,070,000 owner occupiers are estimated to have received tax relief on mortgages in 1990, with an average annual value of 3,150 Hfl.

Subsidies towards company investment   Until recently, investment was stimulated by the Wet Investeringsregeling. Investment in property was subsidized if it was for a new or expanded building for business purposes. However, expenditure on land, including the site on which a subsidized building stood; on buildings for housing or social and cultural activities; and on buildings for rent was excluded from subsidy. With the sale of a subsidized property, part of the subsidy had to be repaid. This subsidy was handled through the tax system, whereby tax due was reduced by the amount of the subsidy. Financial cuts have almost killed off this subsidy. On 28 February 1988, the basic subsidy was reduced from 12.5% to zero. Another part of this subsidy was the possibility of deducting part of the costs of investment from profits, and hence of paying less tax. This was withdrawn on 1 January 1990. All that remains is a bonus for "small-scale investments".

A subsidy is available (*Investeringspremieregeling*) for investment (including property) in certain parts of the country, as part of regional economic policy, as described in Chapter 1.2.

Subsidies for registered monuments   The legislation regulating the protection of buildings and other monuments is described in Chapter 7.1. Under this legislation, two types of subsidy are available from the central government, through the ministry of welfare, public health and culture.

The first is paid towards the capital costs of restoring monuments registered under the Monuments Act. There is a fund (Rijkssubsidieregeling Restauratie Monumenten) of 90 million Hfl per year available for municipalities that have submitted a multi-year programme, containing proposals for restoration work. Individual applications are judged jointly by the national agency and the municipality. The subsidy is for 60%, 70% or 80% of restoration costs, depending on the legal status of the applicant.

The second subsidy is paid towards operating costs (Rijkssubsidieregeling Onderhoud Monumenten). Nationally, 9 million Hfl is available for this. Certain categories of registered monuments are eligible so long as they are in good condition, and normal maintenance (a minimum of 5,000 Hfl, a maximum of 10,000 Hfl) can be subsidized up to 40%. The national agency decides on the application (NWR 1990).

Some provinces and municipalities offer subsidies as well, out of their own funds or the urban renewal fund. Sometimes the subsidies are given to supplement the subsidies from the national agency, sometimes they are based on the municipality's own ordinances.

Grants under the Town and Village Renewal Act The money available under the Town and Village Renewal Act, in the form of the urban renewal fund, has been described in Chapter 3.3. It may be used to subsidize the land development process, housing improvement, registered monuments, etc.

Provision for compensation for loss of value, on the same basis for property as for land, is described in Chapter 3.3.

CHAPTER 8

The process

8.1 Price-setting

Housing

Prices of secondhand, owner-occupied housing are set in the market. There is little to suggest that in the Netherlands the process is different from that in other countries with a free market in owner-occupied housing. There are, however, two peculiarities. One is the transfer tax of 6% on all property sales, including housing (see Ch. 3.3). This decreases mobility within the owner-occupied sector and therefore the speed of adjustment of house prices to changed conditions.

Secondly, statistical analysis of house prices (Janssen 1992) shows that a very large part of the variance in transaction prices in a given year can be "explained" by a few variables. These describe the property itself (age, size, number of rooms, etc.). Adding variables describing the location contributes very little to the explanation. This suggests that location does not have a systematic effect on price. Similar research (hedonic models of house prices) in Britain and the USA, however, finds that location is very important. This confirms the statement made in Chapter 5.2 that the Netherlands is a very uniform country in respect of landed property.

The prices of new owner-occupied housing will be strongly influenced by prices on the secondhand market since the two sources are near substitutes and sales of new housing make up only one-third of all sales, as most houses are supplied from stock (see Ch 9.2). Considering the very great intervention by the public sector in new owner-occupied housing, as supplier of building land, and of subsidies to *premie* and free-sector *premie*-C houses, it is important to ask what the effect of this is on sales prices.

Because new housing is a substitute for secondhand housing, the market value of the former, even when new housing is subsidized, is set in the market for all owner-occupied housing. The subsidies for new housing allow more people, from lower income levels, to buy housing. This will push up prices, including the market value of subsidized housing. The public bodies might want to use their powers to ensure that such houses are sold below their market value. They can do this by imposing price controls on developers of houses receiving an object subsidy (see Ch. 7.3), and then preventing any resulting speculation by the owners by imposing anti-speculation clauses. They can also do it by supplying land below its market value. In order to prevent the resulting development gains being enjoyed by the developer, some municipalities transfer the land to the developer (who will later sell the completed dwelling plus land) in such a way that the developer does not get legal ownership of the land (an ABC-contract). The effect of such practices on sales prices has not yet been researched.

The effect of the quasi-subsidy, whereby mortgage payers receive tax relief on interest payments, is to increase the market price of all houses.

For new social-rented housing, rents are set by a method described in Chapter 7.3. For secondhand rented housing, rents must be set by agreement between landlord and tenant, with the possibility of referring the matter to the rent committee or the court. For rented housing built after 1 July 1989 and with a rent above 750 Hfl a month, rents are set by reference to market levels. Such matters are regulated by the landlord–tenant legislation set out in Chapter 7.1.

Offices

Rental levels for offices supplied for rent (new or secondhand) are set in the market. There is an active market in this sort of space and it is analyzed regularly (see Ch. 9), so the market is transparent.

When offices are supplied for sale (new or secondhand), they can be rental space bought by an investor, or space bought by a final user. In the first case, the price is set by capitalizing the expected income stream. "Best practice" would seem to be to use discounted cashflow techniques. This, however, is not commonly done in the Netherlands. Instead, current net rental income is divided by an initial yield. Negotiations between buyer and seller then centre on the level of that initial yield: if the buyer can be convinced that rents will rise, or that the risks of vacancies or expensive maintenance are low, he will agree to a low initial yield and hence a high price. An analysis of yields used when setting prices is given Chapter 9.3.

In the second case, when offices are bought by the final user, the price is set by reference to capital values on the rental market. This is because the office developer can sell either to a firm for its own use or to an investor as landlord, so he has no reason to sell more cheaply to the one than to the other.

When new offices are built, if they are supplied for rent or sale, then prices are determined as described above. If they are commissioned by their final users, they have no market price, only a cost. Part of that cost is for land. When the municipality supplies that land, it will supply at the same price to the final user as to the speculative developer.

Industrial premises

As stated in Chapter 9.1, by far the majority of industrial space is owned by the final user. A small amount is built speculatively for sale to the final user as, for example, in the case of the Alpha Business Park (Ch. 6.3), and a small amount is rented commercially as, for example, in the Veemarkt case (Ch. 10.3). In the latter case, rental levels are set in the market. The market, however, is not very transparent as it is so small, both absolutely and relative to total volume of industrial space. When industrial premises are sold on the open market, the sales price is set by reference to the rental income, as with offices. When industrial premises are built by the final user, they have no price but a cost. Part of that cost is for the land: cases are known where the municipality has supplied industrial land cheaply to the final user and refused to supply to a commercial developer of industrial property, arguing that the latter would result in the final user paying higher prices (see, for example, the case studies in Needham & Kruijt 1992).

Shops

When shops are supplied for rent (new or secondhand), rental levels for new leases are set in the market. For existing leases, the tenant is strongly protected (see Ch. 7.1). This results in rental levels for existing leases that do not follow changes in supply and demand quickly, which then affects rental levels for new leases. For such reasons, the market in shop leases is not very transparent. The data are analyzed in Chapter 9.2, but they are much fewer than for offices. The lack of transparency is a reason why, in the negotiations over the rental level for a shop, reference is often made to the turnover the retailer will expect to make from the premises. The rent is not set as a percentage of turnover, but turnover is a reference point.

When shops are supplied for sale, the price (both for the investor and the final user) is set by capitalizing the income stream. The practice is to take the current net rental income and to multiply by a "capitalization factor": the inverse of the initial yield (the capitalization factor is often expressed ×10). As with offices, negotiations centre around this: the older the property and the poorer its condition, the lower the capitalization factor and the lower the sales price.

Valuation methods

Properties can be valued for many different purposes. Often, the particular purpose carries its own definition of value. That, combined with the variety of different types of property, leads to many different valuation methods.

Here we concentrate on valuations that aim to assess the market value of a property. But even within this one aim, it is necessary to distinguish between valuations for active and passive purposes. With the former, the property is being offered for sale; with the latter the intention is to assess the value of the property as an asset (e.g. for inheritance tax) but not to buy or sell it. If the property is not actively being offered for sale, it can be subject to various encumbrances (e.g. it is occupied), which lower its value. In that case, the property is valued according to the instructions (i.e. the appropriate definition of value), after which the assessed value is reduced to take account of the encumbrances (e.g. with housing, the rule of thumb is that occupation reduces the value by 40%).

For owner-occupied housing, the method usually applied is comparison: the valuer assesses what the house should bring on the open market on the basis of thorough knowledge of recent sales prices for comparable properties. Valuers have great experience in making such comparisons. Sometimes they try to extend the scope of those comparisons by recording "price-raising" and "price-lowering" factors, and by putting a value on these factors. Then a property to be valued is compared with the most comparable property recently sold, and the latter sales price is adjusted to take account of any differences in price-raising and price-lowering factors. The underlying idea is that the market value of a house is built up of the value of its component factors.

For private rented housing, offices, industrial premises, and shops, the method usually applied is capitalization of the net rental income. This has been described earlier in this section.

Occasionally, properties that are very rarely transacted on the market have to be valued, such as town halls or power stations. In these cases there are no market prices or rents to refer to and the method applied is to estimate the replacement value of the building works (construction cost less depreciation) and, separately, the value of the land (if necessary, for an alternative use that would be possible in that location). Estimating the replacement value of the building works is the method used when valuing for insurance purposes: the value of the land on which the building stands does not have to be assessed, for the land does not have to be insured.

As described in Chapter 7.3, the municipality may levy a local property tax, basing it on the market value of the properties as if they were unencumbered and available for sale (Gemeentewet, Article 273). For this purpose, properties are required to be valued every five years. This is clearly an enormous task, and most municipalities try to systematize it so that, as far as possible, it is not necessary to value each property separately. In most cases, separate valuations will be necessary for shops, offices, industrial premises, and the like.

145

For housing (far and away the most numerous of the properties to be valued), the practice is often as follows. Dwellings are grouped into categories, each category containing similar dwellings in the same general location. For each category separately, one or two "reference points" (i.e. representative dwellings) are chosen. Every five years, these reference points are revalued. The result is then applied to all other dwellings in that category: sometimes the results are applied after adjustment for "price-raising" and "price-lowering" factors. Clearly, this makes periodic revaluations very much easier. Municipalities usually employ outside experts to perform the whole revaluation.

The legislation requires that properties be valued for this tax within bands of 3,000 Hfl: it is common practice to place the property in one or more bands lower than it should be according to the valuation, since this reduces the number of formal objections. And the municipality can still raise the necessary income by fixing the tax rate (rate poundage) correspondingly higher. Local property tax is described more fully in Chapter 3.3.

Arbitration

If agreement cannot be reached about a price or a valuation, it is usually possible to go to arbitration. Most of these possibilities were described in Chapters 4.1 and 7.1. In the case of local property tax, the rules are as follows. If a taxpayer disagrees with his tax assessment, he may submit an objection in writing to the municipal executive or to the head of the finance department. The municipal executive gives a ruling on this objection. If the taxpayer is still not satisfied, he can lodge an appeal with the court (Gerechtshof). The ruling by the court is definite. An appeal can be made to a higher authority only on procedural grounds (Hoge Raad; see Bangma et al. 1990: 110 *et seq.*).

8.2 The actors and their behaviour

The public sector commissions new buildings

Many agencies within the public sector commission buildings from the construction industry for their own use: town halls, civil offices, etc. An alternative way for these agencies to obtain the space they want is to buy or rent it new or secondhand. This way is often followed by central government, which has a special agency for that purpose (see below). The value of work commissioned by the public sector per year is given in Chapter 9.2.

Public agencies can also commission buildings for the use of others. This has happened regularly only in the case of municipalities that ran their own housing associations. A municipality may set up its own association only if it looks as though the private associations cannot cope with local housing needs (Housing Act, Article 61). However, these municipal housing associations now build very little and in many cases they have sold their stock to semi-public housing associations, in line with central government policy. The rapid reduction in their number since 1985 is clear from Table 8.1.

Table 8.1 Number of municipal housing associations.

Number of dwellings owned	1985	1986	1987	1988	1989
0–599	225	191	166	165	
6006–1799	56	59	58	53	
1800–3999	17	16	16	17	
4000–10,000	14	11	12	10	
10,000 and more	5	6	5	5	
total	317	283	257	250	255

Source: Min. VROM, 1990 (c).

Most building works commissioned by the public sector are also financed by that sector (Brouwer 1990: 79). The public sector also finances much building work commissioned by others, granting capital subsidies or loans. The effect of this can be seen in Chapter 9.1. Note, however, that loans from central government for constructing Housing Act dwellings were terminated in 1988.

The semi-public trusts, etc. commission new buildings

Here we are talking of trusts and foundations (*stichtingen*) that do not operate for profit and are closely linked by their statutes to the public administration. In the Netherlands, these provide most education, medical care and social housing. These commission buildings for their own use, often with capital loans or subsidies from the public sector. They can also commission buildings for use by others.

The most important of these semi-public trusts are the housing associations. These are private agencies with full legal rights. They can be officially recognized by the Crown (*toegelaten instellingen*) under Article 59 of the Housing Act only if they are non-profit organizations and engaged solely in housing. It has been estimated (Brouwer 1990: 44) that in 1985 they commissioned 25% by value of all new building works. Information about the number and size of these recognized housing associations is presented in Table 8.2.

Amalgamations have reduced the number and increased the size of housing

Table 8.2 The number and size of recognized housing associations.

Number of dwellings owned	1958	1967	1977	1984	1985	1986	1987	1988	1989
0–599	870	680	421	242	234	213	206	202	
600–1,799	88	171	249	306	303	295	295	284	
1,800–3,999	47	95	160	213	213	225	222	213	
4,000–10,000	⎫			88	93	102	106	107	
	⎬ 27	63	104						
10,000 plus	⎭			10	11	11	12	14	
Total	1,032	1,009	934	859	854	846	841	820	833

Source: Min VROM 1990(c).

associations. The national organizations which represent local housing associations are described in Chapter 4.2.

Others who commission new buildings

This is a very broad category, including, for example, a few owner-occupiers, sports clubs and churches commissioning for their own use. Also, most industrial property and a great part of office construction is commissioned by final users (see Ch. 9.1).

Property developers commission for sale. Therefore they build only those properties for which there is a large and ready market. The property developer sells either to the final user or to the landlord, a property management company or property investment company. Landlords are also interested only in property that can be rented to a large and ready market. The result is that most property developers build either housing for sale or offices (see Ch. 9.1).

Property developers in the Netherlands can be divided into three types. Some (e.g. Amstelland Vastgoed bv, Wilma Vastgoed bv) are attached to large construction firms. An important consideration for these developers is continuity of projects for the mother company. Some (e.g. bv Ontwikkelingsmaatschappij AMRO and MBO Maatschappij voor Bedrijfsobjecten nv) are attached to banks or insurance companies: their aim is to sell the completed project to an investor that will borrow the finance from its mother company. The third type is formed by property developers in the strict sense of the term, such as Blauwhoed Vastgoedontwikkeling bv, Multi Vastgoed. These are independent and follow their own objectives.

Property developers can also commission for use by others. They then manage the property, or work in combination with the final property manager or investor (who may finance the development). Examples may be found in the office sector.

Suppliers of secondhand buildings

Any owner can decide he wants to get rid of his building: he can demolish it and rebuild on the site, or he can sell it. An important consideration is the price he can expect to get if he sells. This will depend on the building (if it was built specially to meet his own needs, it might have a low resale value) and also on the site (it might be a poor building on a valuable site). The buyer could be a new final user, or the existing final user if the landlord sells to a sitting tenant, or a new landlord. There is an active market in secondhand housing and offices, and a less active one in secondhand shops and industrial premises.

The public sector as demander of buildings

There are two special points to be noted about public-sector demand for buildings. Demand for offices from this sector is well known. The importance of various public agencies (such as the public utilities) in the demand for industrial premises is less well known, although between 1983 and 1987 it stood at about $60,000 \text{ m}^2$ per year, equal to about 8% of all demand to buy or rent industrial floorspace (Needham & Kruijt 1992).

The second point is that accommodation for central government and its agencies is the responsibility of the national property agency (Rijksgebouwendienst). Throughout the whole country, this owns or rents, and then manages, buildings for the civil service and also some universities. The most important users include the ministries in Den Haag, the inland revenue, the judiciary, the state police, the prison service, museums, colleges and universities. In 1987, the extent of the buildings thus operated was 6.5 million m^2 gross, with a total value of around 20 billion Hfl. Of these 85% were owned by the agency, the remaining 15% being rented. Under the policy of deregulation and decentralization (see Ch. 2.6), more emphasis is being given to renting instead of owning as this gives greater flexibility, to contracting out various activities, and to a better matching of supply and demand (information from Min. VROM 1988).

Other demanders of buildings

The range of other demanders is huge. Some will want to buy for own use, some will want to rent, and for some there will be a real choice between buying and renting: the latter choice will depend on the relative costs of buying and renting but also on the possibilities of taking out a mortgage and the effect of that on liquidity.

Owners of property for investment

Property management and property investment companies let buildings out to others. They can be divided into three types:

o companies that invest in nothing but landed property;
o the institutional investors, which have money from insurance premiums and pensions to invest, and which invest in all sorts of securities, including landed property;
o all other companies that invest in all sorts of securities, including landed property.

The importance of these companies as providers of property for rent is made clear below. Also, they sometimes act together with a property developer to commission new buildings. They are very important in the market for offices and shops, are becoming increasingly important in the market for industrial property, and for many years have owned and managed a considerable part of the stock of subsidized rented housing. Most is known about the institutional investors. The composition of their portfolios in 1987 is shown in Table 8.3. One can conclude that only a small proportion of their funds is invested in landed property. However, it must be added that the funds available to institutional investors are huge and growing. So their ownership of landed property is, absolutely, huge and growing. Of funds that are so invested, how are they divided between the different types of property? The figures for a few companies are given in Table 8.4.

The institutional investors do not put a large part of their funds into landed property. This share can be compared with that of UK companies. Dutch institutional investors had a smaller share in landed property than their UK counterparts. The figures in Table 8.5 are for 1982.

The construction industry is the real supplier of new buildings, and is described and analyzed in Chapter 1.4.

Table 8.3 The composition of the portfolios of institutional investors.

	Pension funds	Life-assurance companies
	Percentages	
fixed interest securities	81	76
shares	9	9
landed property	9	8
other	1	7
total	100	100

Source: CBS Statistisch Bulletin 31, 10 Sept. 1989

Table 8.4 Ownership of property by selected companies.

	ABP[1]	N.N.[2]	Shell[3]	Bogamij[4]	W.I[5]	Sample[6]
			Percentages			
housing	4.1	47	5	2	1	37.6
offices	53.0	30	85	64	17	35.6
shops	22.5	8	0	30	4	10.1
industrial	20.4	3	10	4	78	(see offices)
offices for own use		9				
other		3				16.7
total	100	100	100	100	100	100

Notes: (1) Algemeen Burgelijk Pensioenfonds, distribution of portfolio floor area, 1988.

(2) Nationale Nederlanden, distribution by value, only those properties in the Netherlands, 1989.

(3) Shell Pensioenfonds, distribution by value, only those properties in the Netherlands, 1989.

(4) Bogamij. This is not an institutional investor but a commercial investment company. Distribution by value, 1989.

(5) West Invest, 1988. This is not an institutional investor but a small property investment company which specializes in industrial property.

(6) Sample survey of institutional investors (O.T.B. 1988)

Table 8.5 Pension fund investment in the UK and the Netherlands.

	Pension funds		Insurance companies	
	Netherlands	UK	Netherlands	UK
		Percentages		
fixed interest	85	26	88	38
securities	3	54	3	37
shares	11	16	8	20
landed property	1	4	1	5
total	100	100	100	100

Source: Breevast, 1984

The property professionals

Property professionals include advisors to clients and the construction industry, such as architects, building surveyors, and structural engineers. There is no reason to think that their effect on the functioning of the property market is anything but neutral.

Another sort of professional advisor are the estate agents who specialize in selling or letting property and the valuers who assess its value. An estate agent provides services as a broker in concluding a contract on the instruction of and on behalf of a person with whom he has no permanent relationship. Estate agents are not allowed to act for both the buyer and the seller in one transaction. If they meet the requisite professional standards, they are sworn in (authorized) by the court (*arrondissement*). In the Netherlands, the professional title of broker in real estate is protected by law.

A valuer estimates the value of a property. A professional valuer works on the instruction of, but independently from, the client. Valuers also can be sworn in by the court (*arrondissement*). The functions of estate agent and valuer can be combined in the same person.

Most brokers are members of the Nederlandse Vereniging van Makelaars (NVM) which has just over 2,000 members (1991) and includes a section for those who want to specialize in commercial real estate. Training for that profession is organized by the professional body itself, at an academic level below that of the universities. There is only one university course in real estate, a part-time post-graduate course at the University of Amsterdam.

Households: buying or renting

Households as actors on the property market have the choice between buying and renting their dwellings. Insight into this choice can be gained by comparing the housing expenditure of renters and of owners. Table 8.6 presents this information in terms of the share of income spent on housing: housing expenditure (gross rent or gross mortgage), minus subsidies received directly (subject subsidies for renters, subsidies for buyers of *premie* housing), plus or minus the effects of tax for the owner occupier (see Ch.

Table 8.6 Housing expenditure of renters and owners.

	1986	1987	1988	1989	1990
			Percentages		
rented	16	16	17	17	17
owner occupied	12	12	11	12	12

Source: Min. VROM, 1990(c).

7.3), expressed as a percentage of disposable household income after tax. Clearly, renters pay a larger part of their income on housing than owners, and that share is increasing.

The actual expenditures (Hfl per month) are shown in the next table, distinguishing between different income groups. For renters, the figures are given in Table 8.7. For owner occupiers, the figures are given in Table 8.8. If these figures are converted into share of income spent on housing, we find

Table 8.7 Housing and other expenditures by renters.

	Minimum[1]		Modal		2 and 3×Modal	
	1986	1989	1986	1989	1986	1989
net rent	285	301	423	435	507	567
gas and electricity	206	158	230	174	239	167
other charges	35	37	34	38	34	37
non-housing expenditure	983	997	1,872	1,938	3,810	3,981
total disposable income	1,508	1,493	2,559	2,584	4,590	4,752
subject subsidy	58	75	7	7	0	0
% receiving subject subsidy	40.9	51.3	6.0	6.6	0	0

Source: Min. VROM, 1990(c), pp. 77, 84

(1) The income groups are, expressed in taxable income ×1000 Hfl

minimum		modal		2 and 3 x modal	
1986	1989	1986	1989	1986	1989
12.5–24	12.5–24	29–35.5	30–37	57–105.5	60.5–111

that, comparing 1986 and 1989, the proportion of renter households paying a large share of income in rent is growing and the proportion of owner households paying a large share of income for purchase is falling (Min. VROM 1990c: 68).

It is quite clear that richer people tend to buy, poorer people to rent. This is most clearly seen when a dwelling is occupied for the first time. Then, in the years that follow, the occupant's income can change without the occupant

Table 8.8 Housing and other expenditures by owner occupiers.

	Minimum[1]		Modal		2 and 3×Modal	
	1986	1989	1986	1989	1986	1989
net acquisition costs	270	270	381	393	470	520
gas and electricity	248	187	256	189	304	208
other charges	45	48	45	48	45	48
non-housing expenditure	1,462	1,490	2,414	2,451	4,645	4,828
total disposable income	2,025	1,995	3,096	3,081	5,464	5,604
tax advantage	83	72	170	149	280	288

For source and notes see table 8.7.

moving. As a result, rented dwellings come to be lived in by some richer people, owner-occupied dwellings by some poorer people. Table 8.9 shows this.

Table 8.9. Income and tenure choice over time

Disposable household income p.a. (Hfl)	First occupants as % of all first occupants in that income group	Occupants as % of all occupants in that income group
	1989	1990
less than 15,000	96% rent	73% rent
25,00 to 30,000 25,00 to 29,000	41% rent	67% rent
more than 45,000 more than 47,000	13% rent	28% rent

Sources: VROM, DGVH, 1991; Ministerie VROM, 1990(c)

Reasons for the behaviour of the actors

Actors on the property market can act from many different motivations, treating property as a good to be produced commercially, as a good to be traded commercially, as an investment good, as a factor of production in a commercial production process, as a factor of production in a non-commercial production process, as an asset, as a consumption good, and as a symbol of status. In this respect, the Netherlands is not significantly different from other industrialized countries.

The public sector as an actor in the property market is motivated by the need for property as a necessary factor of production for supplying public services, and the wish to regulate the actions of others on the property market by controls, grants, etc. In this respect also, the Netherlands is not very different.

What is interesting about public sector involvement in the property market is the reluctance to get involved in supplying building for use by others for schools, hospitals, housing, etc. The reason is the pluralist ideology which finds its origins in the institutional segregation (*verzuiling*) pursued between 1850 and 1950, whereby the provision of such services is left to religious and social groupings; the public sector restricting itself to co-ordination and financing.

CHAPTER 9

The outcome of
the property markets

9.1 Size and distribution of the building stock

An estimate has been made (Table 9.1) of the replacement value of the total stock of buildings, and of how this has grown. The total value is equivalent to about four times the national income. It should be added that this estimate takes no account of the historic value of so many of the older buildings.

Table 9.1 The replacement value of property.

	Dwellings	Other buildings	Civil engineering work	Total	Indexed (1980=100)
(bln Hfl, 1980 prices)					
1950	186	116	103	405	35
1955	202	131	115	448	39
1960	229	154	130	513	44
1965	263	193	157	613	53
1970	326	257	197	780	67
1975	416	320	244	980	85
1980	498	379	280	1156	100
1985	573	427	306	1306	113

Source: Brouwer, 1990, p. 50.

The stock of housing

The total stock of housing, how it has changed, and how it is expected to change in the next years, is shown in Table 9.2. The geographical distribution of that stock is given in Table 9.3. The tenure composition of the total housing stock is shown in Table 9.4.

Small but clear changes can be seen in the eight-year period (1982–90) covered by Table 9.4. There has been an increase in owner-occupation and a decrease in private renting. Rented housing owned by municipalities has also decreased, as they have transferred it to the housing associations.

Table 9.2 The stock of housing, actual and predicted.

(×1,000)	Stock at beginning of year	Additions new construction	Additions other	Withdrawals	Corrections
1979	4,672	86	2	14	
1984	5,178	113	4	12	5
1985	5,289	98	5	10	2
1986	5,384	103	6	12	2
1987	5,483	110	7	12	2
1988	5,589	118	5	13	
1989	5,699	111	4	13	
1990	5,802	101		13	
1995	6,175	81		16	
2000	6,468				

Source: CBS Maandstatistiek Bouwnijverheid.

Table 9.3 The geographical distribution of the housing stock.

(×1,000)	1979	1986	1988	1990
Groningen	195.3	214.8	219.1	224.8
Friesland	196.5	220.7	227.6	234.5
Drenthe	133.8	153.6	159.7	166.3
Overijssel	311.8	342.6	356.0	370.0
Flevoland[1]	18.3	63.5	71.1	79.8
Gelderland	516.0	602.4	629.3	653.3
Utrecht	288.6	341.0	356.3	384.3
N-Holland	821.1	926.0	955.9	990.6
S-Holland	1,098.2	1,252.3	1,290.3	1,318.0
Zeeland	128.5	141.5	146.0	149.4
N-Brabant	627.8	736.2	771.0	809.0
Limburg	355.7	389.4	406.2	422.3
The Netherlands	4,671.6	5,384.1	5,588.6	5,802.4
Amsterdam[2]	301.9	321.0	327.5	335.5
Rotterdam[2]	238.7	268.3	270.6	273.4
The Hague[2]	185.5	193.6	195.8	199.0
Utrecht[2]	82.1	88.9	91.5	93.4

Source: CBS Maandstatistiek Bouwnijverheid.

Notes: (1) Before 1986, Zuidelijke IJsselmeerpolders (Z.IJ.P.).

(2) The dwellings in these cities are included also in the respective provincial totals: Amsterdam in N. Holland, Rotterdam and The Hague in S-Holland, Utrecht in Utrecht.

Table 9.4 The housing stock, by tenure.

Tenure	1982	1986	1990
owner-occupation	42	43	45
rented from a private person	10	8	7
rented from a private organization[1]	7	6	6
rented from a housing association[2]	31	35	36
rented from a municipality or other public agency	8	7	5
rented from others	2	1	1
Total	100	100	100
Absolute (x 1,000)	4,957	5,384	5,802

Source: WBO 1981, WBO 1985/1986, Min. VROM 1990(c).

Notes: (1) Including property management and investment companies.

(2) Including other non-profit organizations.

Table 9.5 Housing tenure by location.

	Social rented[1]	Private rented[2]	Owner occupied	Total	Total abs. (x 1,000)
Groningen	41	17	42	100	224.8
Friesland	39	14	47	100	234.5
Drenthe	36	13	51	100	166.3
Overijssel	37	13	50	100	370.0
Flevoland	45	15	40	100	79.8
Gelderland	34	14	52	100	653.3
Utrecht	34	17	49	100	384.3
N-Holland	36	27	37	100	990.6
S-Holland	38	24	38	100	1,318.0
Zeeland	30	19	51	100	149.4
N-Brabant	34	12	54	100	809.0
Limburg	33	15	52	100	422.3
The Netherlands	36	19	45	100	5,802.4
Amsterdam	41	52	7	100	335.5
Rotterdam	44	39	18	100	273.4
The Hague	36	42	22	100	199.0
Utrecht	42	33	25	100	93.4

Source: DGVH, Provinciaal Tabellenboek Woningmarktinformatie 1990.

Notes: (1) The recognized housing associations.

(2) Private persons, private organizations, non-profit organizations, institutional investors.

Housing tenure varies not only between years, but also between locations. Table 9.5 shows this, giving the situation in 1990. Social rented housing is noticeably more common in the big cities and in Flevoland (the latter partly because that type of housing is relatively new there). Amsterdam stands out as having much more privately rented accommodation and many fewer owner-occupied dwellings than other locations.

The division of the housing stock between single family and multi-family dwellings (flats and maisonettes) was 70% : 30% in 1990. In the four large cities, most of the housing was in multi-family dwellings, ranging from 55% in Utrecht to as high as 87% in Amsterdam (Min. VROM 1990c: 25).

The division of the housing stock by age in 1991 is given in Table 9.6. In 1988, 98% of all dwellings had a bath or shower, 73% had central heating, and the average number of rooms per dwellings was 5.1. The last statistic compares with that for 1899 of 2.75, for 1947 4.79, and for 1985 5.03 (Min. VROM 1990c, Brouwer 1990).

Table 9.6 The housing stock by age and tenure.

	Percentages		
	Owner occupied	Rented	All tenures
Built before 1945	31.5	22.9	26.0
Built 1945–69	27.9	40.1	34.6
Built after 1969	40.6	37.0	39.4
All ages	100	100	100

Source: Min. VROM, 1990(c), p. 30, own recalculations

The quality of the dwellings does not vary much between the sectors, but is more dependent on age. The figures for 1991 in Table 9.7 show this.

The fact that rented dwellings built before 1945 are in a reasonable state of repair can be credited to the urban renewal of the past 20 years. The figures would seem to indicate that rented dwellings were a little poorer in quality than owner-occupied ones. However, the more detailed analysis, which divides rented dwellings into social and privately rented, shows that the quality of social rented dwellings is as high as that of owner-occupied; it is the privately rented dwellings that are in poor condition.

The stock of offices

Not much is known about the total stock of offices. Lukkes (1987) estimates that there are 25 million m² of office floorspace in 50,000 to 60,000 separate buildings; Bak (quoted in De Geus 1990: 6) estimates 26 million m² (leaving aside the difficulty of defining office floorspace). In those premises, 1.2 million to 1.5 million people work, a quarter of the working population.

Table 9.7 The quality of the housing stock.

	Good quality[1]		Reasonable[2] quality		Poor quality[3]		Very poor quality[4]	
	abs. ×1000	%	abs. ×1000	%	abs. ×1000	%	abs. ×1000	%
Owner-occupied, built:								
before 1945	289	34	325	39	157	19	68	8
1945–1969	378	51	292	39	64	9	9	1
after 1969	910	84	159	15	11	1	2	0
total	1,577	59	776	29	232	9	79	3
Rented, built:								
before 1945	270	39	171	24	193	28	61	9
1945–1969	732	57	432	33	83	6	47	4
after 1969	963	78	193	16	78	6	3	0
total	1,965	61	796	25	354	11	111	3
Social rented, built:								
before 1945	96	52	61	33	18	9	12	6
1945–1969	590	59	331	33	45	5	36	3
after 1969	751	79	144	15	61	6	3	0
total	1,437	67	536	25	124	6	51	2
Privately rented, built:								
before 1945	120	29	95	24	151	37	40	10
1945–1969	142	49	101	34	38	13	11	4
after 1969	212	76	49	18	17	6	0	0
total	474	49	245	25	206	21	51	5

Source: Min. VROM, 1990(c).
(1) Repairs cost up to 5% of the value when new.
(2) Repairs cost 5–20% of the value when new.
(3) Repairs cost 20–50% of the value when new.
(4) Repairs cost more than 50% of the value when new, or dwelling cannot be repaired.

Bak also estimated that a half of the stock is owned by the final users, the other half being rented to the final users. Of the half that is rented, about 9 million m² is owned by institutional investors and the rest by other property managers and investors.

The stock of industrial premises

Estimates of the size and value of the stock of industrial property are very difficult to make or find. Blankenstein-Bouwmeester et al. (1984) give a figure of 40 million m², but it is not supported by any argument. Another publication (Min. VROM 1990) gives precisely five times that amount! For manufacturing industry, but not for other productive sectors, official statistics give the value of the stock of land and building at the beginning of 1987 as Hfl 17.5 billion (land) and Hfl 75.3 billion (buildings).

In Chapter 9.2, the question of who commissions new industrial building is discussed. The answer is: almost always the final user. This has consequences for the ownership of the stock. Almost all industrial buildings are owned by their users. Moreover, some of the building commissioned by property developers will subsequently be bought by the final users.

Dutch financial institutions own very little industrial property. When they do invest in real estate, they prefer offices (see Ch. 8.2).

The stock of shops

It has been estimated that there are 6,146 shopping centres (concentrations of six or more shops) of which 2,200 have been comprehensively planned. Of all the retail units, about 63% are in those 6,146 centres, the rest being in smaller clusters or as single shops (Bak 1986). This is all that is known about the size and distribution of the total stock of retail premises.

9.2 Demand for and supply of property

New building

Estimates have been made of the value of the total output of the building industry, divided by type of building work, who commissions it, and who finances it. The estimates for 1978 are given in Table 9.8. It can clearly be seen from Table 9.8 that central government, although not very important in commissioning new works, finances much new building commissioned by others, both private (schools, hospitals, housing associations) and public (provinces, municipalities). Additional information is available for 1950–90 in Brouwer (1990: 43).

More recent estimates have not been made in so much detail. Table 9.9 gives estimates for the total output of the building industry in 1987, with comparable figures for 1978. Particularly striking are the changes in housing, with less spent on new dwellings and more on restoring the housing stock. More detailed information about the composition of the investment in new building is given in Brouwer (1990: 42–3). More detailed information per sector, on who commissions and who finances is given below.

160

Table 9.8 The value of the output of the construction industry, 1978.

(mld Hfl, 1978 prices)	Commissioned by			Total	Financed by		
	central gvt	other gvt bodies	private[1]		central gvt	other gvt bodies	private[1]
New dwellings			14.55	14.55	2.71		11.84
- Housing Act			2.71	2.71	2.71		
- premie-sale			3.93	3.93			3.93
- premie-rent			1.44	1.44			1.44
- free sector			6.47	6.47			6.47
Restoration	0.08	0.29	3.37	3.74	1.37	0.51	1.86
- dwellings			2.27	2.27	1.12	0.10	1.05
- other	0.08	0.29	1.10	1.47	0.25	0.41	0.81
industrial premises			6.20	6.20			6.20
agricultural premises			2.02	2.02			2.02
hospitals			0.73	0.73	0.07	0.21	0.45
schools	0.05	1.14		1.19	0.59	0.60	
for the public administration		0.35	0.62		0.97	0.35	0.62
other buildings			1.22	1.22	0.42	0.09	0.71
civil engineering	1.21	4.28	0.56	6.05	2.22	3.46	0.37
total new construction	1.69	6.33	28.65	36.67	7.73	5.49	23.45
maintenance	0.24	1.21	6.25	7.70	0.24	1.21	6.25
total all construction	1.93	7.54	34.90	44.37	7.97	6.70	29.70

Source: EIB 1980. (1) This includes the semi-public trusts.

Table 9.9 Output of the building industry, 1987.

(bln Hfl)	1987 prices	1978 prices
Dwellings		
new	12.67	14.55
restoration	6.71	2.27
all other new premises	14.22	13.80
civil engineering	5.76	6.05
maintenance	11.26	7.70
total	50.63	44.37

Source: EIB, 1991.

New housing

Table 9.10 presents figures for new house-building, by financial category (see Ch. 7.2) and by tenure. The increase in the proportion of dwellings built for sale in the course of the 1980s can be seen clearly in Table 9.10 (the policy was described in Chapter 2..4), as can the increase in the small number of unsubsidized dwellings built for rent.

Table 9.11 shows that same total of new house-building, but divided by

Table 9.10 New house building by financial category.

(×1,000)	1975		1980	1983	1986		1988	1989
Rented								
for 1 and 2 person households[1]				9.8	5.2			
Housing Act dwellings[1] }	39.9	}	38.7	42.6	31.0	}	40.3	35.6
premie-rent	23.0		10.2	22.0	13.4		8.3	6.7
other	1.2		1.1	1.1	1.0		2.4	3.3
Owner-occupied								
premie-sale	31.4		35.3	30.0	25.8		23.7	21.1
free sector	25.4		27.3	5.6	22.7		43.8	44.5
Total	120.8		113.8	111.1	103.3		118.4	111.2

Source: CBS Statistical Yearbook of the Netherlands.

CBS Statistisch Zakboek. Own recalculations

(1) In 1975, 1980, 1988, 1989, figures for these two categories were published together, not separately.

Table 9.11 New housing by who commissioned it.

(x1,000)	1975		1980	1983	1986
central and local government	4.8		5.7	6.7	3.1
housing associations	41.1		38.8	49.0	37.2
institutional investors[1] ⎤		⎤		8.9	8.3
speculative builders[1] ⎬	74.9	⎬	70.7	33.3	40.3
other private[1] ⎦		⎦		13.2	15.5
total	120.8		113.8	111.1	103.3

Source: CBS Statistical Yearbook of the Netherlands, CBS Statistisch Zakboek.

own recalculations

Note: (1) Before 1980, published together not separately.

who commissions it. The relatively small number of dwellings built by central and local government is clear from Table 9.11. The decrease in the rented sector in the 1980s has meant a decrease in the number commissioned by housing associations. The institutional investors build for rent. The figures conceal the change in their rôle, but from other sources we know that before 1980 they built many more subsidized rented dwellings than they did later. One result is that the share of dwellings in their real estate portfolio fell sharply between 1970 and 1986 (see OTB 1988: 71).

More detail about the types of housing commissioned by the different agencies is available from another source (EIB 1991: 60–2). Building by housing associations is 90% social rented dwellings, and 10% *premie* dwellings for sale. They are very dependent on the housing quotas from central government (see Ch. 4.2). Institutional investors want a secure investment, which can be obtained by building for private renting (47.5% of their construction) and for subsidized (*premie*) renting (52.5%). Most *premie*-rented dwellings are financed by the institutional investors. Speculative builders, both housing developers and general building contractors, build 40% for *premie*-sale, the rest of their output being for the free sector. Of building by individuals, including those who build their own houses, 14.5% is premie-sale, and 85.5% free-sector.

The location of new housing in the past few years is shown in Table 9.12. This makes clear the large amount of new house building that has taken place in the western provinces of N. Holland, S. Holland and Utrecht. Under present policies, the ribbon of cities in N. Brabant (Breda, Tilburg, Eindhoven, 's-Hertogenbosch) and the cities in Gelderland (Ede–Wageningen, Arnhem, Nijmegen) should accommodate more new house building (see Ch. 1.2).

New offices

The annual value of production broken down by who commissions it, is given in Table 9.13 by the analysis of building permits granted.

This source also provides data for the analysis presented in Table 9.14 of the total floorspace covered by the permits granted. Offices commissioned by property developers (not for their own use) are included in the category "trade and transport" in Table 9.14. Other offices in that category will, however, be commissioned by the final user, as will most of the offices in the other two categories. It has been estimated (Lukkes 1987: 10) that between a third and two-fifths of all new offices are built to be rented to others.

From the same source (building permits), the average building costs per square metre are analyzed in Table 9.15. Table 9.16 gives some figures about where that new office floorspace was built.

Table 9.12 The location of new housing.

(x1,000)	1985	1986	1987	1988	1989
Groningen	1.8	2.0	2.5	2.6	3.1
Friesland	3.1	3.9	3.4	3.4	3.6
Drenthe	3.7	2.9	3.1	3.5	3.5
Overijssel	6.6	7.3	6.5	7.7	6.9
Flevoland[1]	3.2	3.2	4.2	4.5	4.4
Gelderland	11.2	13.6	13.1	13.2	11.4
Utrecht	6.5	7.7	7.7	9.0	9.1
N-Holland	16.2	15.2	17.5	17.8	19.0
S-Holland	21.9	20.6	24.3	25.6	21.8
Zeeland	1.4	2.0	2.2	1.9	2.0
N-Brabant	15.8	16.3	18.3	20.5	19.1
Limburg	6.7	8.5	7.3	8.7	7.2
The Netherlands	98.1	103.3	110.1	118.4	111.2
Amsterdam	4.3	3.8	5.7	5.2	4.3
Rotterdam	3.2	3.0	4.2	4.9	3.7
The Hague	2.5	2.0	1.8	3.4	1.5
Utrecht	0.5	1.3	1.3	1.1	0.8
Growth towns and centres[2]	16.8	14.4	15.1	17.5	17.3

Source: Min. VROM, 1990(c), p. 54.
Notes: (1) Before 1 January 1986, the Zuidelijke IJsselmeerpolder.
(2) See Ch. 1, applied for by:1982 1984 1986 1988 1989.

Table 9.13 Value of office production per year.

Applied for by: (mln Hfl)	1982	1984	1986	1988	1989
manufacturing and agriculture	108	208	307	331	430
trade and transport	627	854	846	1,660	1,829
public and semi-public	238	199	424	309	403
total	973	1,261	1,577	2,300	2,689

Source: VGM 11 (1990) p. 30.

Table 9.14 New office floorspace by source of application.

Applied for by:	1982	1984	1986	1988	1989
			percentages		
manufacturing and agriculture	15	18	21	14	16
trade and transport	65	69	57	75	68
public and semi-public	20	13	22	11	15
total	100	100	100	100	100

Source: VGM 11 (1990) p. 30.

Table 9.15 Building costs of new offices.

Applied for by: (Hfl per m^2)	1982	1984	1986	1988	1989
manufacturing and agriculture	766	950	1,190	1,217	1,204
trade and transport	1,056	1,031	1,202	1,108	1,237
public and semi-public	1,301	1,284	1,565	1,386	1,207
total	1,060	1,049	1,279	1,154	1,239

Source: VGM 11 (1990) p. 30, own recalculations

Table 9.16 The location of new office floorspace.

(×1000m^2)	1981	1982	1983	1984	1985	1986	1987	1988	1989	yearly average 1981–88
Amsterdam, region[1]	61.0	72.0	125.0	97.0	114.0	33.0	222.0	77.0	205.0	100.0
The Hague, region[2]	29.5	50.0	90.0	107.0	127.0	67.0	31.0	50.0	-	69.0
Rotterdam, region[3]	43.0	3.0	8.5	16.5	32.0	26.0	55.0	88.0	-	34.0
Utrecht, region[4]	33.0	18.8	99.0	20.0	51.0	46.0	50.0	66.0	-	48.0

Source: De Geus, 1990, p. 29.

(1) Incl. Diemen, Amstelveen, Badhoevedorp.

(2) Incl. Rijswijk, Voorburg, Voorschoten, Leidschendam.

(3) Incl. Schiedam, Capelle a/d IJssel

(4) Incl. Maarsen

The above are the regions used by the journal VastGoedMarkt for analyzing transactions.

The dominance of Amsterdam among the four big cities is clear from Table 9.16. Comparable figures for total production nationally are not available.

New industrial premises

The construction of industrial premises can be analyzed in the same way as for offices, namely from building permits. The value is given in Table 9.17. From the same source (building permits), the average building costs per square metre have been calculated in Table 9.18. The figures in Table 9.18 indicate also who commissions the building of industrial premises. It will be

Table 9.17 The value of output of industrial premises.

Applied for by:	1982	1984	1986	1988	1989
			mln Hfl		
Production space without offices:					
manufacturing	82	128	142	213	298
trade and transport	100	155	286	313	311
property developers[1]			35	16	36
other[2]	88	73	87	77	111
Total	270	356	550	619	755
production space with offices, built by property developers[1]			77	152	222

Source: VGM 13 (1990) P. 33.

Notes: (1) Before 1986, applications by property developers were included in "manufacturing" and in "trade and transport".

(2) Including the public sector and agriculture

Table 9.20 Applications for permission to develop retail premises.

Applied for by:	1980	1987	1988	1989	1990
			percentages		
institutional investor	6	4	2	2	4
property developer and other speculative builders	34	30	20	39	25
retailers	60	66	77	59	71
total	100	100	100	100	100

Source: VGM 12 (1990) p. 34.

seen immediately that property developers are much less important here than in the market for new offices, at least with respect to traditional industrial premises. The rapid growth of combined production/office space is striking and is the subject of the case study in Chapter 6.3. This growth is possibly the cause of the rapid rise in building costs per square metre for "other" industrial premises: combined production/office space costs more to build than traditional factory or warehouse space.

The value of industrial premises built by property developers can be estimated from another source. Of all the industrial property bought or rented on the open market in the period 1983-7, only about 70,000 m² per year (10% of the total) was new (Hogenbirk et al. 1988). If this cost 681 Hfl per m² to build (see above) total construction cost was 48 million Hfl. This is the same order of magnitude as the average for 1986 and 1987 given in Table 9.18. For more details see Needham & Kruijt (1992).

An indication of where those industrial premises are being built is given by the following figures. Analysis of the value of buildings, excluding land, showed that concentrations of industrial property acquired (bought and commissioned) by manufacturing industry were to be found in Twente (3.5%), Arnhem/Nijmegen (13.5%), Utrecht (5.7%), greater Amsterdam (4.3%), greater Rotterdam (Rijnmond; 4.6%), north-east of North-Brabant (4.1%), west and south-east of North-Brabant (16.7%), and southern Limburg (8.2%) (CBS 1988). An analysis of industrial property acquired on the open market (bought and rented) between 1983 and 1987 shows a concentration around Amsterdam, Rotterdam, Utrecht, Eindhoven, 's-Hertogenbosch, Twente, the Flevopolders and the Haarlemmermeerpolder (Hogenbirk et al. 1988). Both studies give a somewhat similar picture of geographical distribution.

New shops

The analysis of building permits granted for retail premises is shown in Table 9.19. The analysis in Table 9.19 compares new building and restoration: this is because of the huge sums spent on re-equipping existing shops and rehabilitating existing shopping centres.

Table 9.20 breaks down the total according to who applied for permits to develop retail premises. According to the figures in Table 9.20, retailers themselves commission most of the total building works. This is not what would have been expected from other figures, which show that most new shopping space is commissioned by institutional investors (VGM 13 1986: 25). No explanation for this discrepancy is available. No information is available about the location of the new shops.

For all new premises other than dwellings, Table 9.21 analyzes occupancy types for new premises, but only for those premises offered for sale or rent

Table 9.19 The value of new retail premises.

	1982	1984	1986	1988	1989
			mln Hfl		
new building	375	273	300	406	546
restoration and rebuilding	102	132	151	169	163
total	477	405	451	575	709
price per m² (Hfl)	793	652	671	676	728

Source: VGM 12 (1990) p. 34.

Table 9.18 The building costs of industrial premises.

Applied for by:	1982	1984	1986	1988	1989
			Hfl per m²		
manufacturing	336	312	329	374	402
trade and transport	400	315	350	372	396
other	332	403	558	592	681
total	356	329	365	391	425

Source: VGM 13 (1990), p. 33.

Table 9.21 Occupiers of new premises by industrial group.

Industrial group	1983–87	1988–89	1989–90
		percentages	
manufacturing and public utilities	11.4	7	5.4
construction and installation	1.4	1	4.5
trade and repairs	6.6	3	5.9
transport and storage	3.6	1	2.4
communications	7.4	4	6.3
banking and insurance	11.7	12	7.3
commercial services, financial	6.8	11	7.8
commercial services, other	8.9	14	17.4
computer sector	12.7	14	9.8
public administration	13.1	15	16.7
education	6.4	6	7.3
other	10.0	11	9.2

Source: Zadelhoff, 1990, p. 8.

on the open market. New premises commissioned by the final user or by an investor for renting are not included. The relative importance of the occupiers is indicated by the percentage of the floorspace they take. The decline in the importance of manufacturing is probably significant, as is the decline in banking and insurance, and the growth in commercial services.

All properties (new and existing)
Housing The number of dwellings and their type bought and sold between 1985 and 1989 is given in Table 9.22. This reflects the number and type of dwellings built, as indicated above. The number of dwellings sold per year out of the existing stock can be compared with the size of the stock of owner-occupied dwellings (in 1986, 43% of 5,384,000 = 2,315,000). About 5.6% change hands each year. Chapter 1.3 showed that the "turnover rate" for owner-occupied dwellings was a little over 5%.

Table 9.22 The number of dwellings bought and sold.

	1985	1986	1987	1988	1989
New dwellings:					
premie-sale A	18,400	18,600	19,200	19,700	17,800
premie-sale B	8,700	8,200	6,100	3,000	700
premie-sale C	5,800	11,400	11,700	12,400	11,000
free sector	15,900	14,700	13,700	19,000	26,300
total	48,800	52,900	50,700	54,600	55,800
existing dwellings	109,700	128,800	121,100	121,000	115,000
total	158,500	181,700	172,800	175,600	170,800

Source: Min. VROM, 1990(c), p. 79.

If supply and demand do not match each other, the result can be price adjustments, waiting lists or vacancies. Changes in house prices are described in Chapter 9.3. Waiting lists or vacancies for owner-occupied dwellings can be measured by the number of days between a dwelling being offered for sale and being bought. Some figures for this are presented in Table 9.23.

The figures in Table 9.23 give a very clear picture of changes in the market for owner-occupied dwellings. In the second half of the 1970s it was a sellers' market. This changed rapidly in the first half of the 1980s, with great oversupply. Since then, the market has slowly come back into equilibrium. The same pattern can be seen in changes in house prices (Ch. 9.3).

Vacancies are registered, and the figures for the whole country are given in Table 9.24. This shows that the total number of vacancies is growing

Table 9.23 The time taken to sell-owner occupied housing.

(number of days)			
1976: 55	1980: 107	1984: 138	1988: 113
1977: 53	1981: 113	1985: 153	1989: 101
1978: 75	1982: 115	1986: 149	1990: 102
1979: 100	1983: 123	1987: 127	

Source: MCC and NVM. Reworked by J.Janssen, Nijmegen.

Table 9.24 Housing vacancies.

	Unsuitable for habitation	Suitable for habitation				Total
		already been occupied		never been occupied		
		vacant less than 4 month	vacant more than 4 month	vacant less than 4 month	vacant more than 4 month	
1979	9,300	36,600	51,300	2,300	2,700	93,000
1985	13,100	55,600	67,200	3,700	6,000	126,400
1986	15,200	51,100	66,100	2,700	4,400	124,300
1987	14,200	52,200	66,900	2,900	3,900	125,900
1988	16,100	51,900	64,000	4,000	3,900	123,800
1989	16,100	51,400	68,900	4,600	5,100	130,000

Source: CBS Maandstatistiek Bouwnijverheid, in Min. VROM, 1990(c), p. 23.

slowly: one of the reasons could be that the demand comes from households which are getting smaller, and that the supply (the existing housing stock) cannot react quickly to this. Those vacancies, expressed as a percentage of the total housing stock, are given in Table 9.25 for each province. The high vacancy rate in Flevoland is striking. Towns in this polder area were built and expanded for people moving out of the Amsterdam region, but it would appear that the overspill policy is not fully successful.

Figures for housing vacancies do not distinguish between the different types of dwelling standing empty. In 1991, the CBS began to collect that additional information, which will be available from 1992.

Offices The volume of transactions on the office market is defined as "the total area that in a certain period is sold or rented on the open market" (De Geus 1990). The figures are presented in Table 9.26. It will be seen that the four large cities accounted for about 60% of all transactions by volume, and

Table 9.25 Housing vacancies in each province

	1979	1985	1986	1987	1988	1989
			percentage of housing stock			
Groningen	2.7	2.3	2.3	2.4	2.5	2.8
Friesland	2.1	2.3	2.5	2.6	2.4	2.2
Drenthe	1.6	1.8	1.7	1.9	1.9	2.0
Overijssel	1.7	2.0	2.2	2.1	2.1	2.0
Flevoland[1]	0.9	4.3	4.0	4.4	4.3	4.1
Gelderland	1.6	1.9	1.8	1.8	1.9	2.0
Utrecht	2.0	2.0	1.8	1.7	1.7	2.1
N-Holland	1.9	2.6	2.4	2.2	2.1	2.2
S-Holland	2.9	3.4	3.1	3.0	2.7	2.6
Zeeland	3.1	2.9	3.1	3.6	3.7	3.9
N-Brabant	1.6	1.5	1.5	1.5	1.6	1.7
Limburg	1.6	1.8	1.9	2.0	2.1	2.3
The Netherlands	2.1	2.4	2.3	2.3	2.2	2.3
Amsterdam	–	4.0	3.5	3.0	2.7	2.6
Rotterdam	–	4.4	4.0	3.8	3.4	3.0
The Hague	–	5.2	5.1	4.9	3.6	3.7
Utrecht	–	2.0	2.3	2.2	1.8	3.0

Source: CBS, Maandstatistiek Bouwnijverheid.
(1) Before 1 January 1986, Zuidelijk IJsselmeerpolders.

Table 9.26 The volume of transactions in the office market.

	1980–84	1985–89	1989	1990
		m² x1,000		
Amsterdam region	93	221	343	247
The Hague region	85	108	79	166
Rotterdam region	45	123	176	132
Utrecht region	34	89	98	161
elsewhere	158	398	558	675
The Netherlands	415	939	1254	1381

Source: De Geus, 1990, p. 24; VGM 1 (1991) p. 3.
Note: For the regions, see Ch. 9.3.

that the volume of transactions has been increasing over the past 10 years.

The amount of office floorspace offered for rent or sale on the open market in the course of a year is presented in Table 9.27. The volume offered on the market in the course of a year can be compared with the volume transacted in that year. The comparison must be interpreted with caution, but in this case it is clear that supply is greater than transactions, a result consistent with the existence of much empty office space.

Table 9.27 Office floorspace offered for sale or rent.

(x1,000 m²)	1980	1982	1984	1986	1988	1989	1990
Amsterdam region	261	378	368	360	352	333	540
The Hague region	136	233	363	238	234	225	229
Rotterdam region	152	147	217	279	316	287	307
Utrecht region	62	57	129	110	104	100	125
elsewhere	567	637	665	546	667	826	951
The Netherlands	1,178	1,452	1,742	1,533	1,673	1,771	2,152

Source: De Geus, 1990, p. 27.
Note: For the regions, see Ch. 9.3.

Table 9.28 shows how the volume of vacant office space has grown in the past 10 years. That such vacancies are not only absolutely high, but relatively high also, is shown by Table 9.29 (for 1989). An explanation for high vacancy rates in the four big cities is given by the figures in Table 9.30 of new office floorspace put onto the market.

Table 9.28 The volume of vacant office space.

(m²)	Yearly average	
	1981–84	1985–89
Amsterdam region	165,000	247,000
The Hague region	85,000	179,000
Rotterdam region	93,000	157,000
Utrecht region	32,500	54,200

Source: De Geus, 1990, p. 26.
Note: for the regions, see Ch. 9.3.

Clearly, demand has not been able to absorb the large increase in supply. Moreover, in the four cities so many new offices are in the pipeline and so much land for office building is being offered by the municipalities that no end to oversupply can be expected in the near future. Table 9.31 presents for the four big city regions the total stock of offices, the current plans for new

Table 9.29 Area and percentage of vacant office floorspace.

	Total area office floorspace m^2	Area vacant office floorspace m^2	Vacant as % total
Amsterdam region	4,000,000	260,000	6.5
The Hague region	3,400,000	170,000	5.0
Rotterdam region	2,700,000	167,000	6.2
Utrecht region	1,300,000	48,500	3.7

Source: De Geus, 1990, p. 29.
Note: for the regions, see Ch. 9.3.

Table 9.30 New office floorspace, 1981–88.

	Annual average between 1981 and 1988 m^2
Amsterdam region	100,000
The Hague region	69,000
Rotterdam region	34,000
Utrecht region	48,000

Source: De Geus, 1990, p. 29.
Note: for the regions, see Ch. 9.3.

Table 9.31 Office stock, plans and occupation.

	Stock, 1989 m^2	Current plans absolute m^2	As % of stock	Average net increase in occupation 1981–88 m^2	Current plans sufficient for Years
Amsterdam region	4,000,000	2,658,000	66	83,000	32
The Hague region	3,400,000	1,070,000	31	41,500	26
Rotterdam region	2,700,000	1,754,000	65	23,500	75
Utrecht region	1,300,000	970,000	75	43,500	22

Source: De Geus, 1990, p. 36.
Note: for the regions, see Ch. 9.3.

offices (absolute and as a proportion of total stock), the average yearly net increase in office floorspace occupied, and the number of years it will take before current plans are fully occupied if that net increase continues unchanged. Clearly, if current plans are realized in the next 5 to 10 years, there will be a massive oversupply of offices in the cities.

Industrial premises   Little research has been done on the volume of transactions in industrial premises. Table 9.32 presents the results of an analysis of transactions reported in *VastGoedMarkt*. Only transactions of properties with more than 750 m² are included. Moreover, it is probable that for reasons of confidentiality not all transactions will have been reported.

Table 9.32 indicates that the volume of transactions increased in the last few years of the 1980s, except in the Amsterdam region.

Information is available about the amount of industrial floorspace offered for sale or rent on the open market in the course of a year, how it has changed, and where it is offered. This information is presented in Table 9.33. In 1990, more than half of the supply was in the Randstad. Other

Table 9.32 The volume of transactions of industrial premises.

	1985	1986	1987	1988	1989	1990	1985-90 ave. p.a.	1985–90 rented/bought
Amsterdam region[1]	54	37	60	66	67	56	57	77% / 23%
N-Holland province[2]	40	51	92	47	61	102	65	64% / 36%
Rotterdam region[3]	47	56	59	65	66	121	69	70% / 30%
S-Holland province[4]	48	139	54	76	125	108	92	57% / 43%
Utrecht province	62	77	70	128	123	138	100	72% / 28%
total Randstad	251	360	335	382	442	525	383	67% / 33%

Source: Rouwenhorst, 1991.

(1) Incl. Amstelveen, Diemen, Duivendrecht. (2) Excl. Amsterdam region.

(3) Incl. Capelle a/d IJssel. (4) Excl. Rotterdam region.

Table 9.33 The amount of industrial floorspace offered for rent or sale.

	1981	1983	1985	1986	1987	1988	1989	1990
				x1,000 m²				
in the Randstad	1,378	1,631	1,207	917	800	854	913	1,073
elsewhere	1,138	1,489	1,050	987	793	868	1,012	1,044
total	2,516	3,120	2,257	1,904	1,593	1,722	1,925	2,117

Source: VGM 1 (1991) p. 5.

concentrations in 1990 were: Enschede/Hengelo (100,000 m^2), the city of Utrecht (95,000 m^2), Zoetermeer (40,000 m^2), Venlo (60,000 m^2) and the cities of N. Brabant (together, 195,000 m^2).

In the Randstad, the volume of transactions recorded was less than half of the volume of supply recorded, for every year since 1985. One would expect, therefore, considerable vacancies. But nothing systematic is known about them in this sector. The oversupply in industrial land is described in Chapter 5.3.

Shops For shops, nothing is known about the volume of transactions, but figures about the floorspace offered for sale or rent on the open market in the course of a year are available. These figures are presented in Table 9.34. The share of the Randstad in the whole is less than 50%, smaller than for industrial premises and even less than for offices. One would, however, expect a more even locational distribution of shops as they must be near to customers. Nothing systematic is known about vacancies in this sector.

Table 9.34 Retail space offered for sale or rent.

	1981	1983	1985	1986	1987	1988	1989	1990
				x1,000 m^2				
in the Randstad	202	225	245	165	117	116	162	162
elsewhere	298	290	292	210	221	207	223	277
total	500	515	537	375	338	323	358	439

Source: VGM 1 (1991), p. 7.

All commercial premises Table 9.35 gives an overview of where commercial premises are being offered for sale or rent on the open market, for 1 December 1990. Only the larger premises have been analyzed. Once again, the dominance of the Randstad is clear, especially in the office market.

9.3 Prices and rents

Housing

Prices of owner-occupied housing All property sales, including housing, have to be registered with the public cadastral service (see Ch. 3.1). Until 1985, these prices were analyzed by the CBS and published in Maandstatistiek Bouwnijverheid. Since then, the cadastral service itself (KADOR) has analyzed and published house prices, but its statistics exclude sales below 40,000 Hfl and above 400,000 Hfl. Figures from both sources are presented in Table 9.36: they should be compared with caution. Most striking is the very rapid rise between 1970 and 1978, followed by a very rapid fall up to 1982, and a slow rise since then, by which the top prices of 1978 are being approached but have not yet been reached.

Table 9.35 Commercial premises offered for sale or rent.

	Offices in premises bigger than 500 m²	Industrial/ warehouses in premises bigger than 750 m²	Shops/ showrooms in premises bigger than 200 m²
Groningen	38,000	60,000	16,000
Friesland	19,000	17,000	22,000
Drenthe	24,000	43,000	16,000
Overijssel	65,000	180,000	36,000
Gelderland	59,000	161,000	34,000
Utrecht	283,000	218,000	20,000
Flevoland	30,000	33,000	7,000
N-Holland (excl. Amsterdam region)	165,000	182,000	51,000
Amsterdam (excl. Amstelveen, Duivendrecht and Diemen)	537,000	181,000	21,000
S-Holland (excl. The Hague region and Rotterdam region)	171,000	275,000	29,000
The Hague (incl. Leidschendam, Rijswijk, Scheveningen, Voorburg, Wassenaar)	229,000	64,000	18,000
Rotterdam (incl. Schiedam and Capelle a/d IJssel)	307,000	131,000	18,000
Zeeland	3,000	8,000	6,000
N-Brabant	193,000	435,000	84,000
Limburg	29,000	129,000	61,000
Total	2,152,000	2,117,000	439,000
N-Holland/S-Holland Utrecht/Flevoland (incl. the large cities)	1,722,000	1,084,000	164,000

Source: VGM 1 (1991) p. 11.

Table 9.36 House prices.

(Hfl)	Detached houses[1]	Other single family houses[1]	Flats and apartments[1]	All housing[1]	All housing[2]
1970	117,200	42,300	49,600	49,100	
1971	114,900	46,800	50,500	55,600	
1972	123,300	52,500	49,200	60,600	
1973	143,200	60,300	52,800	68,400	
1974					
1975					
1976	231,800	96,600	76,900	108,800	
1977	315,700	126,700	100,300	143,500	
1978	381,900	146,300	114,300	166,200	155,500
1979	381,000	144,400	115,700	161,900	145,500
1980	354,600	135,000	97,900	151,400	131,100
1981	286,200	116,400	88,600	138,900	121,400
1982	248,600	102,800	79,200		118,700
1983	243,200	105,700	78,900		118,700
1984	250,000	106,400	80,900		118,500
1985	247,200	106,100	77,600		121,800
1986					124,900
1987					129,400
1988					132,100
1989					137,700
1990[3]					143,200

Sources: See below.

(1) CBS Maandstatistiek Bouwnijverheid.

(2) KADOR 1991.

(3) This is the average price in May. For all other years, the price given is the unweighted average of average prices in November and May.

Table 9.37 compares changes in house prices with changes in general price levels and in income per person. Comparing the beginning and end of the 20-year period 1970–90, owner-occupied housing has risen in price faster than the general price level but not as fast as the level of income per person. However, in the middle of that period, house prices had risen very much faster than both inflation and personal incomes.

Figures of sales price by region are also available. However, the regions are not the standard regions used by others, but the cadastral service's own regions (for the description of these regions see KADOR 1991). A selection of these regions is used for this analysis. The prices are for transactions in the month of November (for 1991, in February). Transactions at prices below 40,000 Hfl and above 400,000 Hfl are not included.

Table 9.37 Index of changes in house prices and the cost of living.

	House prices[1]	Cost of[2] living	Level of[3] income p.p.
1970	100	100	100
1975	221.6[4]	151.8	149.5
1980	308.4	204.8	220.3
1985	248.1	250.5	274.2
1987	263.5	250.0	282.7
1989	280.4	250.5	314.1
1990	291.6	256.1	337.5

Sources: See below.

(1) 1980, CBS, Maandstatistiek Bouwnijverheid, thereafter KADOR 1991. Note that these two sources are not strictly comparable.

(2) CBS, Statistisch Zakboek.

(3) see Ch1.2

(4) The value is not available for 1975, this is the value for 1976.

Table 9.38 presents the figures in such a way that house-price differences between the regions are clear. It also shows the stability of that regional differentiation.

It can be seen that prices are not highest in the Amsterdam region, the country's capital, but in the Utrecht region. Nor are prices consistently lower

Table 9.38 Regional house price differentials.

	Amsterdam abs. Hfl	index	Arnhem	Breda	Roer-mond	Rotter-dam	The Hague	Utrecht
1977	163,135	100	94	90	86	74	81	119
1978	168,667	100	105	99	73	81	82	112
1979	150,640	100	108	92	104	85	84	104
1980	130,410	100	107	95	102	88	80	110
1981	110,024	100	112	107	117	103	96	115
1982	114,888	100	107	104	117	97	92	110
1983	117,369	100	104	104	104	98	83	117
1984	121,922	100	100	96	100	92	85	107
1985	122,126	100	102	99	102	92	87	107
1986	127,338	100	95	100	101	96	79	113
1987	126,345	100	105	99	106	96	90	115
1988	133,164	100	104	99	98	97	86	106
1989	128,235	100	113	108	116	94	96	123
1990	148,768	100	95	94	99	87	86	108
1991	142,954	100	99	96	114	93	92	111

Source: KADOR, 1991, own recalculations

the farther one moves away from the Randstad. The two big cities of Rotterdam and Den Haag have consistently the lowest prices. This is probably because they have a large proportion of old (and therefore cheap), dwellings in their stocks of owner-occupied housings, smaller plot sizes, and higher proportions of multi-family dwellings. For new dwellings, prices in the west are higher than elsewhere, as is shown below. Table 9.39 presents the same figures but in such a way that the change in prices can be compared in the different regions. Prices were higher in 1991 than in 1977 in only two regions.

Table 9.39 Regional house price differentials over time.

	Amsterdam	Arnhem	Breda	Roermond	Rotterdam	The Hague	Utrecht
				(Hfl and index)			
1977	163,135	153,691	146,599	140,113	120,918	132,076	193,822
1977	100	100	100	100	100	100	100
1978	103	116	114	88	113	104	97
1979	92	106	94	112	106	96	81
1980	80	91	85	95	95	79	74
1981	67	80	81	92	94	80	66
1982	70	80	82	96	92	80	65
1983	72	80	83	87	95	74	71
1984	75	79	80	87	93	78	67
1985	75	81	82	89	93	81	68
1986	78	79	87	92	101	76	75
1987	77	87	86	95	100	86	75
1988	82	90	90	93	106	86	73
1989	79	95	94	106	100	93	81
1990	91	92	96	105	108	97	83
1991	88	92	93	117	109	99	82

Source: KADOR, 1991, own recalculations

Finally, and from a different source, Table 9.40 compare the sales prices of new dwellings in the four standard regions. For new single-family houses, sales prices are highest in the west and lowest in the east, but once again the very small geographical variations are striking. For "other single-family houses" in particular, there is no more than a 4% difference between the price in the cheapest region and the most expensive region.

There is a final aspect of house prices to investigate: the share of land costs in the total price. This can be most reliably calculated by using the sales prices of new houses and by comparing these with the plot prices given in Chapter 5.3. This is done in Table 9.41.

In one sense, the results of Table 9.41 are clear: for *premie*-sale houses, plot prices are a stable 16% of sales price; for free-sector owner-occupied houses, plot prices as a share of sales price have been rising rapidly. Interpreting those results is, however, more difficult. For *premie*-sale houses, what we see is the result of government policy, not market forces. That same percentage (cost of land between 13% and 17% of the cost of land plus construction) was applied by the central government in the period 1953–86 as a way of setting the maximum land costs for Housing Act dwellings (RIGO 1989). For free-sector owner-occupied houses, municipalities have been demanding higher plot prices while sales prices have been falling; the fact that the municipalities have been able to realize these plot prices means that the market will bear them.

Table 9.40 The sale price of new dwellings, by regions.

	Detached houses	Other single family houses	Flats and apartments
	Hfl	Hfl	Hfl
North			
1984, 1st quarter	210,400	136,600	114,700
1985 1st quarter	197,500	137,300	141,200
1985 2nd quarter	222,800	138,300	177,600
East			
1984 1st quarter	200,200	135,700	130,500
1985 1st quarter	224,500	135,300	141,300
1985 2nd quarter	205,400	135,500	157,000
South			
1984 1st quarter	215,700	140,300	139,700
1985 1st quarter	214,400	139,000	139,300
1985 2nd quarter	213,600	140,500	139,400
West			
1984 1st quarter	235,400	139,100	144,300
1985 1st quarter	244,400	139,200	135,200
1985 2nd quarter	245,800	141,200	139,700
The Netherlands			
1984 1st quarter	222,200	138,400	140,300
1985 1st quarter	231,800	138,000	136,500
1985 2nd quarter	231,100	139,500	142,600

Source: CBS Maandstatistiek Bouwnijverheid, 1988, Table 14.9 H.

Table 9.41 House prices and plot prices.

	1983 Hfl	1984 Hfl	1985 Hfl	1986 Hfl	1988 Hfl	1989 Hfl
House prices						
premie – sale A	143,000	140,000	139,000	134,000	—	—
premie – sale B	161,000	153,000	151,000	146,000	—	—
all premie-sale	145,000	143,000	143,000	139,000	—	—
free sector premie – C	n/a	n/a	—	128,000	—	—
all free sector	242,000	219,000	183,000	180,000	—	—
all new housing	159,000	153,000	156,000	155,000	175,000	180,000
Plot prices						
premie – sale	23,800	23,500	23,400	23,800	23,200	—
free sector premie – C	23,200	26,000	27,100	27,100	26,700	—
other free sector	36,700	41,200	48,300	49,100	42,600	—
Plot price as						
% sales price:						
premie – sale (A+B)	16	16	16	17	—	—
premie – sale C	—	—	—	21	—	—
other free sector	15	19	26	27	—	—

Source: sales prices 1983 to 1986, Willems-Schreuder 1987,
sales prices 1988 and 1989, NRO 1989 and 1990; plot prices, see Ch5.3.
Notes: The plot prices for "other free sector" have been compared with sales
prices for "all free sector", so the results should be interpreted with care.

Price of rented housing Housing rents are given in Table 9.42. The figures
are for "bare rents" (*kale huur*) i.e. the price paid for use of the property,
excluding service costs, heating, etc. The average rent is given, with the
distribution between rent classes. Rents are given for 1 July, the date on
which rent increases come into force. The increase in average rent can be
seen, and the diminishing share of cheap dwellings and the corresponding
increase in expensive ones.

Increases in housing rents are compared with increases in the rents for
production space and with inflation in Table 9.43. It is clear that rents for
housing have risen faster than rents for other buildings and faster than
inflation. The cause can be found partly in government policy: obligatory
rent rises are specified during the first five years for rented dwellings built

Table 9.42 Housing rents

rent class (Hfl per month)	1979	1986	1989
		percentages	
below 99	5.1	0.7	0.1
100–149	11.8	1.6	0.5
150–199	24.8	4.3	1.3
200–249	18.4	9.7	3.2
250–299	14.0	14.9	7.5
300–349	9.4	13.7	13.1
350–399	7.1	11.6	13.3
400–449	4.3	11.2	12.8
450–499	2.2	9.8	11.0
500–549	1.3	8.0	10.6
550–599	0.7	6.1	8.6
600–649	0.5	3.6	7.4
650–699	0.1	2.2	4.5
700–749	0.1	1.0	2.6
above 750	0.2	1.5	3.5
total	100.0	100.0	100.0
average (Hfl)	244	389	435

Source: CBS Woonkostenenquete, in Min. VROM, 1990(c), p. 70.

Table 9.43 Inflation in housing rents and production space.

	Dwellings	Production space	Inflation
		percentages	
in the period			
1974–1978	8.0	6.7	7.4
1978–1982	7.6	2.8	5.8
1982–1986	5.2	1.6	2.1

Source: Brouwer, 1990, p. 65.

with a subsidy, and rents of other housing can be raised so that all dwellings of the same quality carry the same rent (*huurharmonisatie*, see also Ch. 7.1).

Obviously, rents will vary with the quality of dwellings, as shown in Table 9.44. The variation in rent is, in most cases, as would be expected. Perhaps some explanation is needed of the variation with year of construction. If rent varied with quality, we would not necessarily expect newer dwellings to have

Table 9.44 Rents and the quality of dwellings.

	1979	1986	1989
	Hfl per month		
Year of construction			
before 1931	158	254	297
1931–1959	191	294	333
1960–1969	250	375	416
1970–1974	355	487	529
1975 and later	–	501	533
No. of rooms			
1–3	204	335	377
4	223	369	416
5	244	391	438
6	286	449	502
7 and more	302	466	515
Type			
single family	252	418	468
multi family	235	354	396
Housing Act dwelling			
yes	235	380	424
no	254	402	453
Average rent	244	389	435

Source: CBS Woonkostenenquête, in Min. VROM,
1990(c). p. 71.

higher rents, because newer dwellings are not always of better quality. We have seen, however, that rents for dwellings built with a government subsidy are determined for the first few years, partly according to total costs: it is probably partly because newer dwellings have higher construction costs that they have higher rents.

The variation of rents with location is shown in Table 9.45. The geographical variation in rents is perhaps an unexpected one. Rented dwellings are scarcest in the four big cities, so why are rents there lower than average? And why are they higher than average in Flevoland, which has such a high vacancy rate? The answer lies partly in the relationship established above: rents are lower for older dwellings, and the housing stock is older than average in the four big cities, and younger in Flevoland. We have seen in Chapter 5.3 that the market relationship between demand and

Table 9.45 Rent variations by location.

	1979	1986	1989
		Hfl per month	
Groningen	234	375	416
Friesland	234	378	420
Drenthe	234	388	443
Overijssel	227[1]	376[1]	426
Flevoland	–	467	509
Gelderland	256[2]	410[2]	447
Utrecht	259	412	458
N-Holland	235	370	410
S-Holland	246	387	434
Zeeland	249	393	436
N-Brabant	251	401	450
Limburg	252	407	453
The Netherlands	244	389	435
Amsterdam	217	322	359
Rotterdam	205	324	370
The Hague	233	360	401
Utrecht	229	364	407

Source: CBS, Woonkostenenquête, in Min. VROM,
1990(c), p. 69.
Notes: (1) Incl. Noord-Oostpolder.
(2) Incl. Zuidelijk IJsselmeerpolder

supply is not allowed to influence plot prices for new social-rented housing. Clearly, it is not allowed to influence rental levels either.

Offices

The sales price of office buildings is determined by the combination of rents and yields, as explained in Chapter 8.1.

An analysis has been made of all offices offered for rent and of the asking rents. From this, weighted averages have been calculated. The results are presented in Table 9.46. Rents have been rising very slightly (faster than inflation only in Amsterdam) and are higher in Amsterdam than elsewhere.

An analysis is also available of rent is in the four big cities (see Table 9.47), but only for the "top locations", not the average for transactions in all locations. The "top-10" office locations are listed in Table 9.48, along with the rents in May 1991. Offices in Amsterdam-Zuid are particularly popular for financial and commercial services, but it was not until 1985 that annual rents there rose above 300 Hfl per m², for the World Trade Center. Now a new building (Twin Tower) being constructed next to that is expected to fetch 500 Hfl. It seems as though Amsterdam is catching up on cities such

Table 9.46 Office rents.

(Hfl per m² p.a.)	1983	1984	1985	1986	1987	1988	1989	1989/1983
Amsterdam region	213	229	219	205	189	221	239	112.2%
The Hague region	207	223	212	216	214	221	206	99.5%
Rotterdam region	189	185	177	176	192	184	198	104.8%
Utrecht region	220	205	193	208	219	224	221	100.5%
cost of living								107%

Source: De Geus, 1990, p. 30. Note for the regions see Ch.9.3.

Table 9.47 Office rents in the "top locations".

(Hfl per m² pa)	1980	1982	1984	1986	1988	1989	1990
Amsterdam	275	275	280	295	350	375	430
The Hague	280	300	300	285	280	300	325
Rotterdam	225	230	250	260	270	275	290
Utrecht	225	235	245	255	255	270	285

Source: Arthur Andersen, 1991.

Table 9.48 Office rents in the "top ten locations".

		Hfl per m² pa		
1.	Amsterdam – South	450	–	500
2.	The Hague – centre	300	–	350
3.	Amsterdam – centre	300		
4.	Amstelveen (aggl. A'dam)	300		
5.	Rotterdam – Weena	270	–	300
6.	Utrecht – centre	270	–	300
7.	Utrecht – Rijnsweerd	265		
8.	Rotterdam – Brainpark	260		
9.	Amsterdam – South East	250		
10.	Hoofddorp (aggl. A'dam)	250		

Source: VIB, Vastned en Zadelhoff, in Volkskrant 6 May 1991.

as Milan, Madrid and Frankfurt. Nevertheless, considered internationally, office prices are still low. This is because demand is spread between the four big cities and their agglomerations, and supply is also spread, with every municipality trying to sell land to office developers (see Ch. 5.2).

The limited concentration of office property is illustrated further with the figures in Table 9.49 for office rents (Zadelhoff 1990). There is surprisingly little geographical variation between the rents for "top locations" and rents in fairly peripheral areas such as Almelo and Nijmegen. The excessively low rents in Lelystad, in Flevoland, are caused by a huge oversupply.

Information about office yields is less complete than about rents. In particular, the yields actually enjoyed by investors on offices acquired some years ago are a commercial secret. Nevertheless, it would appear that investors would have done better to have put their money into government bonds than into offices in the period (a good example of "reverse yields").

What is known are the expected initial yields on new or renovated offices being offered for sale or rent. These are presented in Table 9.50 (with, for comparison, yields on five-year government bonds) for offices in A1 locations. A supplementary analysis of initial yields (1989–90) is given in Table 9.51.

Industrial premises

There is only a small active market in industrial property, in which prices can be set. Figures show that rents were at "dump level" for several years, but are now beginning to grow, especially in Amsterdam and Rotterdam (Needham & Kruijt 1992).

Absolute levels of annual rents in 1989–90 were on average 80 Hfl per m^2, with little variation around that average, from as low as 40 Hfl in the peripheral regions of the north and east, to around 100 Hfl in parts of the Randstad and 160 Hfl at Schiphol (Zadelhoff 1990). Table 9.52 gives rents in more geographical detail. For industrial premises, just as for offices and housing, it is striking how little geographical variation in rents there is.

Changes in the period 1985–90 have been analyzed for regions in the Randstad. The analysis was conducted on transactions of over 750 m^2 reported in *VastGoedMarkt* for regions defined as in Chapter 9.2, and is shown in Table 9.53. These price rises are surprising considering that between 1985 and 1990, much more space was offered for sale or rent than was taken up. The analysis of price rises for production space in the whole country for the period 1978–82 shows much slower growth (see above). An explanation for the high rises in Table 9.53 could be that the figures analyzed are from transactions, and that the quality of the premises being traded has risen.

Table 9.49 Office rent concentration.

City (Hfl per m²)	Area	Rent	City	Area	Rent
Almelo		170–190		Diezeborgh	215
Amersfoort	station area	225		Helftheuvel	215
	Stadsring	215		Herven/	
Amstelveen		250–300		Business Park	195
Amsterdam	Centre	200–395		Kantoren/	
	Zuidoost	225–250		de Herven	215
Buitenveldert	250–285			Hoofddorp	225–240
	Station Zuid	300–425	Houten	Molenzoom	210
	West	215–250	Leeuwarden		150–225
	Teleport/Sloterdijk	225–240	Leiden		175–235
	Oud Zuid	250–425	Leidschendam		180–210
Apeldoorn	town centre	185–205	Lelystad	town centre	125
Arnhem	Velperweg	215		centre North	115
	Zuid	190–210		Lelycentre	100
	Station	250	Maastricht	MECC	200–215
Baarn	De drie eiken	225		Stationplein	185–195
Breda	Haagse Beemden	195	Nieuwegein	Kantoren Centre	230
	Prinsenlaan	195–215		Zuidstedeweg	230
	Bedrijvenpark Breda	165	Nijmegen	centre	185
Bunnik	Entrada Bunnik	220		Lindenholt	190
Capelle	Hoofdweg area	195–210	Rijswijk		175–225
The Hague	near main station	300	Rosmalen	Brabantpoort	220
	Lange Voorhout	250–325	Rotterdam	Oosterhof	230
	Parkstraat	275–295		Blaak	240–270
	Congresgeb'gebied	290		Brainpark	240–255
	Binckhorstlaan	210		Weena	250–280
	Kerketuinen/			Zuidplein	210
	Zichtenberg	140–170	Schiedam		185–210
Deventer	town centre	190–220	Utrecht	Hoog Catharijne	250
Diemen		220–240		"La Vie"	275
Dordrecht	centre	195–210		Daelse Kwint	295
Eindhoven	Eindje	250		Rijnsweerd	250–260
Fellenoord	Boschdijkplein	210–225		Kanaleneiland	240–260
	Mariënhage	240		Kaap Hoorndreef	200
	Schimmelt	250		Lunetten	215
	Eindhoven Airport	200		Oude stad	220
	Science park	190–210		Hojel complex	325
	High Tech Parc	180–195	Venlo		180
Enschede	centre	185–210	Woerden		185–210
	Business/		Zeist	Boulevard and/	
	Science park	160–180		surroundings	220
Groningen	Corpus den Hoorn	200		Utrechtseweg	230
	Martini Trade Park	210		Lenteleven	200
	Stationsweg/			Mooi Zeist	225
	Herewegzone	225	Zoetermeer		180–210
Hengelo		180–210	Zwolle	along A 28	200–210
's-Hertogenbosch	Pettelaar Park	210–235		Oosterenk	205
	Soetelieve Noord	235–245		near centre	210
	Hooghe Herven	210		town centre	200

Source: Zadelhoff, 1990.

Table 9.50 Office yields.

	1980	1982	1984	1986	1988	1989	av.1980–89
initial yield net %	6	7.5	7.75	7.25	6.7	6.4	7
initial yield gross %	6.5	8.0	8.25	7.75	7.2	6.9	7.5
5-year govt bonds	10.13	9.91	8.11	6.36	6.11	7.20	8.1

Source: VGM kerncijfers, Zadelhoff, ABN. In: De Geus, 1990.

Table 9.51 Analysis of office yields, 1989–90.

	Lowest initial yield		Highest initial yield	
	gross	net	gross	net
Offices in the Randstad				
A1 locations	7.0	6.6	7.5	6.8
other locations	7.8–8.0	7.3	9.0	7.75
All commercial property				
Outside the Randstad				
the best locations	7.5–8.0	7.0–7.3	8.25	7.75
other locations	7.8–8.25	7.2–7.6	10.5	8.5

Source: Zadelhoff, 1990.

Table 9.52 Industrial property rents.

City (Hfl per m^2 per yr)	Area	Rent	City	Area	Rent
Almelo		40–45	IJsselstijn		70
Almere		60–80	Leidschendam		70–90
Alphen a/d Rijn		60–85	Lelystad		45–55
Amersfoort		60–85	Maarssen		85–95
Amstelveen		100–120	Maastricht	Randwijck	85–100
Amsterdam/				Beatrixhaven	65–95
Zuidoost		110		Bosscherveld	60–70
Apeldoorn		55–70	Nieuwegein		75–90
Arnhem		60–90	Nieuw Vennep		90–100
Barendrecht	Dierenstein	90–105	Nijmegen		45–70
Breda		50–90	Rijswijk		70–90
Breukelen		80	Roermond		45–70
Capelle a/d			Rotterdam	Noordwest-	
IJssel	Hoofdweggebied	85–105		bedrijven	95–115
	Hoofdplaats	125		Stadionweg	90–105
Culemborg		60–75		Ommoord	90–105
The Hague	Kerketuinen	75–100		Goudsesingel	125
	Binckhorstlaan	85–115		Waalhaven	75–95
Deventer		50–65	Schiphol	(incl.customs	
Diemen		90–100		facilities)	160–180
Dordrecht		75–	Sittard		50–75
Eindhoven		55–90	Tilburg		50–85
Enschede		60–75	Uithoorn		85–95
Groningen	Algemeen	40–60	Utrecht	Lageweide	80–100
	Driebond	80–		Overvecht	75–100
Heerenveen		40–50	Vianen		65–95
Heerlen		65–95	Woerden		65–80
Hengelo		40–50	Zeist		75–85
's-Hertogenbosch		55–95	Zoetermeer		70–110
Hoofddorp	De Noord	90–100	Zwolle	along A 28	55–85
Houten		80			

Source: Zadelhoff, 1990.

Table 9.53 Changes in industrial rents.

	1990 (1985=100)
Amsterdam region	131
N-Holland province[1]	135
Rotterdam region	120
S-Holland province[2]	157
Utrecht province	162
all Randstad	133

Source: Rouwenhorst, 1991.

Notes: (1) Excl. Amsterdam region.

(2) Excl. Rotterdam region.

Table 9.54 Industrial rents and yields in the main cities.

	1989			1990	
	2nd quarter	3rd quarter	4th quarter	1st quarter	2nd quarter
Rents (Hfl per m² pa)					
Amsterdam	70–100	75–110	75–110	75–130	75–140
The Hague	75–110	75–110	80–110	80–110	80–110
Rotterdam	75–110	75–110	75–110	75–110	75–110
Utrecht	80–110	80–115	80–120	80–120	80–130
initial yields (%) all four cities	7.5–8.0	8.0–9.0	7.0–8.0	7.0–8.0	7.0–8.0

Source: Jones Lang Wootton, 1990.

Table 9.55 Additional information on industrial property yields.

	lowest initial yield		highest initial yield	
	gross	net	gross	net
production space in the Randstad				
the best locations	8.25	7.25	8.50	7.75
other locations	9.0	7.75	10.5	8.5
production/office space outside the Randstad				
the best locations	8.75	7.75	10.0	8.75
other locations	9.75	8.5	11.5	9.0
Hi-tech space				
in the Randstad	7.75	7.2	9.0	8.25
elsewhere	8.25	7.5	9.25	8.5

Source: Zadelhoff 1990.

Table 9.56 Retail rental levels.

City	Area	Rent (Hfl per m^2 pa)
Alkmaar	Langestraat	1,000–1,400
Amersfoort	Langestraat/	
	Utrechtsestraat	1,000–1,300
Amsterdam	Kalverstraat	1,400–1,800
	P.C. Hooftstraat	1,000–1,200
Apeldoorn	Hoofdstraat	1,000–1,400
Arnhem	Ketelstraat/	
	Vijzelstraat	1,200–1,500
Assen	Kruisstraat	600–700
	Mercuriusplein	700–800
Breda	Ginnikenstraat/	
	Karrestraat/	
	Eindstraat	1,000–1,600
Den Bosch	Hinthamerstraat/	1,000–1,300
	Pensmarkt/	
	Hogesteenweg	1,300–1,600
The Hague	Spuistraat	1,200–1,600
	Venestraat/	
	Vlamingstraat	1,000–1,200
Dordrecht	Sarisgang/	
	Begijnenhof/	
	Statenplein	1,000–1,200
Eindhoven	Demer/Rechtestraat	1,400–1,600
Enschede	Kalanderstraat	1,000–1,200
Groningen	Heerestraat	1,400–1,600
Haarlem	Grote Houtstraat	1,200–1,400
Heerlen	Promenade/	
	Saroleastraat	900–1,100
Hilversum	Kerkstraat	1,000–1,300
Leeuwarden	Nieuwstad(zuid)	1,100–1,300
Leiden	Haarlemmerstraat/	
	Donkersteeg	1,000–1,300
Maastricht	Grote Staat/	
	Kleine Staat	1,400–1,800
Middelburg	Lange Delft	900–1,100
Nijmegen	Broerstraat/	1,200–1,400
	Burchtstraat	1,000–1,200
Rotterdam	Lijnbaan	1,200–1,800
	Hoogstraat	1,200–1,600
Tilburg	Heuvelstraat	1,100–1,400
Utrecht	Hoog Catharijne	1,200–1,800
	L. Elisabethstraat	1,200–1,500
Venlo	Vleesstraat	1,000–1,200
Zwolle	Diezerstraat	1,200–1,400

Source: Zadelhoff, 1991.

Changes in rent in the more recent past, yields and changes in yields, are shown in Table 9.54. There is supplementary information on yields, showing differences between locations and types of industrial property. This is given in Table 9.55.

Shops

Some absolute rental levels are given for 1991 in Table 9.56. The figures are for a standard shop unit of $100\,m^2$ to $120\,m^2$ with a frontage of $6\,m$, and in an A1-location. It is noteworthy that the highest rents are not in the four big cities, and that rents in those cities are no higher than rents in smaller towns such as Eindhoven and Nijmegen.

How rents have changed in the recent past, also yields and how those have changed, are shown in Table 9.57.

Table 9.57 Changes in retail rents and yields over time.

	1989			1990	
	2nd quarter	3rd quarter	4th quarter	1st quarter	2nd quarter
rents (Hfl per m² pa)					
Amsterdam	1,600	1,650	1,650	1,700	1,750
The Hague	1,550	1,550	1,600	1,600	1,650
Rotterdam	1,600	1,650	1,650	1,700	1,750
Utrecht	1,350	1,350	1,400	1,400	1,450
initial yields (%) all four cities	6.25	6.0–6.4	6.0–6.5	6.0–6.5	6.0–6.5

Source: Jones Lang Wootton, 1990.

9.4 Speculation in property

Pure speculation will occur when somebody buys property in the hope that its value will increase without any further action. If someone succeeds in buying and selling within a period of three months, no transaction tax is levied (see Ch. 3.2).

No tax is specifically levied on development gains, although capital gains are taxed under company income tax (see Ch. 3.3). There is no development gains tax in the Netherlands, because nearly all land development is managed and controlled by the public administration (the municipalities), and the balance of demand and supply of land and properties is mostly in equilibrium, so prices have not risen enough to reward speculation on any scale.

Risks associated with property development and ownership

When a developer of property builds speculatively (without knowing who will buy the completed property) he has to take account of the risk of vacancy. There are other sorts of risk in property ownership for investment: lowering of rental value and rental growth, lowering of sales prices, risk of a bad tenant, risk of higher management costs, etc. Some investments will be more affected by uncertainty than others and hence risk analysis will be more relevant to some investments than others.

In the Netherlands the risk in property investment is lower than the risk in owning shares but higher than the risk in owning government bonds.

CHAPTER 10

Case studies

10.1 Brainpark, Rotterdam: construction of offices in a growing city

Context of the development

Rotterdam is a city with about 580,000 inhabitants in the south wing of the Randstad (see general location map). It is associated with manufacturing industry and the activities of the biggest port in the world. Office activities, especially commercial services, are therefore poorly represented. Recently, however, much office floorspace has been built there as data in Chapter 9.2 show, although less than in the Amsterdam region, and office rents are lower in Rotterdam than in Amsterdam or Den Haag (see Ch. 9.3). The municipality of Rotterdam wants the city to develop into an internationally oriented service centre and is actively encouraging office development. Its plans are so ambitious that they could cause a massive oversupply (see Ch. 9.2).

Municipal policy distinguishes between several types of office location: office districts in the centre, dispersed within the central area, peripheral transport nodes, etc. This case study is about an office estate, named Brainpark, at a peripheral transport node. The location, 3 km from the city centre, is served by the metro with park-and-ride facilities, buses, trams, and a junction to the A16 motorway (one of the busiest in the country). It is intended for smaller free-standing offices in a parkland setting. It will be seen that this location satisfies the criteria set for an A- or B-location specified by central government for employment locations (see Ch. 2.1), as well as having excellent access to a motorway.

In 1985 the municipal council decided to investigate the possibility of developing this location for high quality businesses. It was hoped to attract middle-sized offices (1,500–3,500 m²), high-tech activities which use office-like buildings, and firms with links to Rotterdam's Erasmus University, which together with a business school, Hogere Economisch School, is next to the site. It was this latter consideration that led to the name Brainpark.

It was decided to develop an area of 18 ha that was already in possession of the municipality. Brainpark I (see Fig. 10.1), as it was called, was a success and an adjacent site is to be developed as Brainpark II. This second site has an area of 8.5 ha, of which 5.8 ha is suitable for building. It was used partly for grazing, partly for glasshouse horticulture.

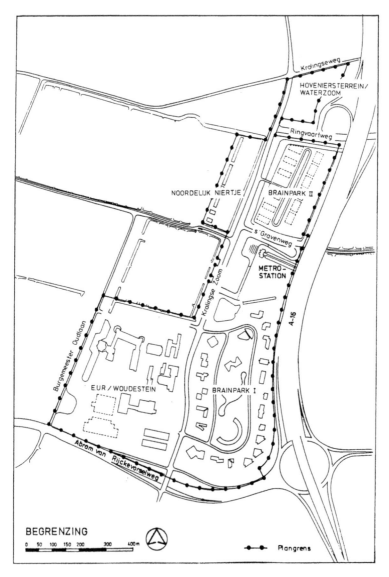

Figure 10.1 Brainpark office estate, Rotterdam.

The content of the plan

Brainpark I falls within a land-use plan dating from 1957 (Uitbreidingsplan in hoofdzaak Kralingen) which designated this site for housing. Brainpark II falls within a land-use plan dating from 1974 (Bestemmingsplan voor gronden ten oosten van Kralingse Zoom tussen Hertoginnenweg en 's-Graven-hage) which designated the site for housing. Along the boundary with the motorway, social rented flats were to be built, in the form of a high wall to act as a noise barrier that would protect private housing to be built behind it. As recently as 1983, the designation of this general location for housing was confirmed. But under the influence of the economic recession, the emphasis came to be placed on employment following the council's decision of 1985. However, it was not until 1989 that a start was made on a new land-use plan, and a draft version was published in July 1991.

So far, the municipality has steered and controlled the development using two sets of powers. Under public law, it used the anticipatory procedure of the Physical Planning Act (previously Article 19, now Article 29) to allow development contrary to the valid land-use plan "in anticipation of" the content of a new land-use plan (see Ch. 3.1). And under private law, the municipality used its powers as owner of the land. Together, these powers allow a tight control over the development.

Brainpark I should offer 19 plots of 3,000–5,000 m² each around the edge of the site, and 2 plots of 12,000 m² in the centre (see Fig. 10.1). A total 70,000 m² of office and production floorspace and a hotel with 80 rooms should be built on these plots. This should create employment for 3,200.

A great deal of attention is paid in the plan to the layout of the "office park". It specifies that no more than 66% of a plot may be built upon, and that the first 10–15m of a plot is a zone on which no building or parking may take place. There is one parking space for every 50 m² of floorspace, and all parking for a unit is to be within its own plot. The maximum building height on the peripheral plots is four storeys, and on the central plots eight storeys, so that the central buildings are visible from afar. The hotel, on a plot in the south west corner (see Fig. 10.1) may be up to 20 storeys high. The layout of each plot must be approved and every firm must make a "green plan" so that the park-like landscape can be safeguarded. The firms themselves must set up an association with the task of maintaining (with municipal money) public and private open spaces.

Brainpark II should offer 12 plots, ranging from 2,400 m² to 4,200 m², totalling more than 30,000 m² of office space. This is to be developed according to a "total urban design plan". The buildings are to line a broad avenue, each building being surrounded by planting. Although each building will be different, together they must make up an architectural unity. Maximum building height on the eastern plots (along the motorway) is to be five to six storeys, on the western plots two to four storeys. Around 650 parking spaces will be provided, all within the plots.

Plan implementation

The land for Brainpark I was already owned by the municipality, which had begun making it suitable for building by raising its level by several metres by pumping sand onto it. The land for Brainpark II was also municipally owned, but was still being used for agriculture and horticulture. Those firms had to be bought out and the buildings demolished. The land was 5m lower than Brainpark I, and the difference in levels is being retained. The land is now being serviced. No figures have been made available for acquisition or servicing costs.

Land is disposed of leasehold, and most developers have chosen the option of paying a lump-sum premium for 50 years. This was calculated per square metre of gross floorspace, not per square metre of land. The price is 650 Hfl if the ground floor of the building is to be used for offices; 450 Hfl if it is to be used for showrooms or production. The municipality receives no subsidy towards the land development process, and is itself not subsidizing it. On the contrary, it expects to make a profit on the project.

For Brainpark I, a site manager was appointed to maintain good contacts between developers and the municipality, and there has been intensive promotion of the project. This has contributed to the site's rapid development.

The plots have been disposed of to property developers and to firms building for own use. Within the first year, options had been taken on all the plots (in total 22: 19 peripheral and 3 in the centre, a small change from the plan). Building started in May 1987. In 1988 7,000 m^2 was constructed; in 1989, 36,000 m^2; in 1990, another 26,000 m^2, thus completing Brainpark I.

In order to avoid new offices standing empty, the municipality insisted that a building be 50% pre-let before construction could start. One result has been that four of the buildings were commissioned by firms which themselves moved into a part of them.

In 1989, there were 31 firms on Brainpark I whose activities included insurance, real estate, tax advice, consulting engineering, advertising, computing, other commercial services, manufacturing and construction, trade, transport, and semi-public and other non-commercial services. There is also an 18-storey hotel with 196 rooms.

There are very few high-tech firms, contrary to the aims of the plan. Another aim that has not been realized is functional links with the university: these have not (or not yet) arisen, nor were they a reason why firms chose this location. Facilities mentioned in the first planning stages were a central antenna and an optic-fibre cable for telecommunications, which have not yet been provided, and possibly never will be. Potential occupants of Brainpark II were asked if they wanted such a cable, but said that it was not important for them. The association to maintain the public and private open spaces has not yet been set up.

Annual rents are around 240–50 Hfl per square metre, only a little below city centre rents of 270 Hfl (see Ch. 9.2). Rents in Rotterdam are, however, still below those in Amsterdam and Den Haag.

The firms were asked in 1989 what they thought about Brainpark I. They valued its good reputation, prestige, location, ample parking, and good accessibility. They criticized the absence of facilities such as shops and a post office for lunchtime shopping, and the somewhat unco-ordinated appearance of the estate, which could have been avoided had there been an architectural master plan.

The development of Brainpark II is being tackled differently. A "total urban design plan" has been made, and the whole project has been placed in the hands of three property developers (Blauwhoed, MultiVastgoed, Burginvest) which have each taken on four plots. These developers are working together to achieve "unity in diversity". As in stage I, conditions are being imposed to safeguard against too much space being provided too quickly. The condition is that none of the three developers, each building up to $11,000\,m^2$, may continue building if more than $3,500\,m^2$ of his space is standing vacant.

Conclusions

Clearly, Brainpark I has been a success. All the plots were disposed of and developed quickly. That experience led to the decision to go ahead with Brainpark II.

Not all the aims set down in the planning for Brainpark I have been realized. However, a high-quality business park has been developed and the firms there seem to be satisfied. A disappointment is the unco-ordinated appearance of the whole estate. The municipality has learnt from this and is tackling Brainpark II differently. It is not only that urban design is being controlled more strictly: the municipality has learnt that the demand is so strong that it can be more selective in granting permissions.

The property developers are satisfied with the commercial results achieved with Brainpark I, otherwise they would not be willing to develop the second stage. The project is also a financial success for the municipality, which has made a profit on the land development.

One can criticize the planning procedures followed, but there is no evidence that anyone was harmed by them.

10.2 De Meenthe, Leeuwarden:
Urban renewal in a medium-sized town:

Context of the development

Leeuwarden is in the north east of the Netherlands and is the capital of the province of Friesland (see general location map). It has 85,000 residents. On the northern edge of the city and along the River Dokkumer Ee is a housing development, De Bilgaard. It consists of nine complexes, one of which, De Meenthe, is the subject of this case study (see Fig. 10.2).

De Meenthe was built in 1967–8 and consists of 418 dwellings. These were made up of 179 single-family houses, 209 dwellings in flats and 30 split-level dwellings in flats. The 30 split-level dwellings were made into 15 large flats. The dwellings in flats were in blocks: block (S 12) of 12 storeys with 64 dwellings, one block (S 8) of 8 storeys with 100 dwellings and one block (S 5) of 5 storeys with 60 dwellings.

The block S5 has no lift but is connected by a pedestrian bridge with S8, which (along with S 12) does have a lift. So residents of S5 had to get to the higher storeys via S8. The galleries of all the flats in De Meenthe, unlike elsewhere in De Bilgaard, are enclosed.

The whole complex is fairly isolated and is reached by only one access road. The design of the neighbourhood is large scale and monotonous, and it was not a pleasant place to live in. The owner and landlord is the housing association (*stichting*) De Leeuwarderadeel.

The first problems in De Meenthe became apparent soon after its completion. In 1974 and 1975 vacancies arose as families moved out to housing in the urban fringe. Their place was taken by outsiders and by one- and two-person households.

At the end of the 1970s, the three housing associations owning split-level flats in De Bilgaard asked the National Woningraad (see Ch. 4.2) to investigate living conditions in these complexes. The finding was that people seeking a dwelling in Leeuwarden deliberately avoided a dwelling, particularly a flat, in De Bilgaard and, therefore, in De Meenthe. The researchers recommended improvements such as building works, more maintenance and management, and a different lettings policy. The housing association drew up an improvement plan at the beginning of the 1980s, but it was rejected by the province because it did not meet the conditions for receiving a subsidy.

The complex continued to attract attention from, among others, the University of Groningen and the working party De Meenthe, composed of residents. Meanwhile problems worsened. The municipal council gave priority to improving dwellings built before 1939, thus ignoring De Meenthe. And the housing association De Leeuwarderadeel thought that living

198

conditions could be adequately improved by small-scale works.

As a result the vacancy rate (average per year) increased steadily until 1986: in 1983 it was 21%; 1984, 37%; 1985, 43%; and 1986, 51%. The turnover rate (average per year) showed a similar trend: 1983, 50%; 1984, 59%; 1985, 79%; and 1986, 69% (Het plan van aanpak, 1987). The improvement in the vacancy rate in 1986 is a reflection of a similar trend in the whole country, not of improvements within De Meenthe.

The content of the plan

At the beginning of 1986, the planning group De Meenthe was set up, consisting of representatives from the housing association, the municipality, and the working party De Meenthe (see above). The municipality asked this group to draw up a "plan of attack", which should present recommendations for concrete action to tackle the problem. At the same time, certain management activities, such as maintenance of the planting, refuse disposal and lettings policy, were intensified.

The housing association had wanted to demolish the block S8, where the problems were worst and the vacancies highest. However, the executive committee of the municipality wanted to explore other possibilities first. One was that one of the other local housing corporations might take over that block, but they were prepared to do this only on condition that it was demolished.

On 30 November 1987 the Municipal Council accepted the "plan of attack". There should be a last attempt to avoid writing-off S8. One possibility was to empty and re-let it under special conditions. Another possibility was selling the block: the trust that provided student accommodation was interested.

If the block were to be emptied, 4,000 Hfl would be given to each tenant to cover removal costs, using money from the urban renewal fund. Further, S8 should be disconnected from S5, which would mean that S5 would need its own lift. A selective lettings policy should be introduced. And 200,000-Hfl would be set aside by the municipality to cover possible losses should the housing association decide, after two years of trying to rent the flats, to demolish them (that decision could be taken if the vacancy rate was higher than 50%).

Four options were discussed with the trust for student accommodation:
○ retaining block S8, with substantial alterations paid for under the subsidy system for new housing (see Ch. 7.3);
○ retaining block S8, improving it with subsidies under the housing improvement scheme;
○ demolishing block S8 and rebuilding it under the subsidy system for student accommodation;

○ demolishing block S8, as the housing association had initially proposed. On 13 February 1989, the last of the four options was chosen, because the other options were financially not feasible. Moreover, it would be inadvisable to delay much longer because block S8 had by then been empty for two years, harming the residential environment in the whole estate. In order to demolish the block, permission had to be obtained from the secretary of state for housing to withdraw the building from residential use (see Ch. 7.1).

Plan implementation

Demolition of block S8 began in June 1989, at a total cost of 3.904 million Hfl. Of this, writing off the unredeemed loan from central government (see Ch. 7.2) totalled 3.384 million Hfl and demolition costs 520,000 Hfl.

The cost was financed by a loan to the housing association De Leeuwarderadeel from five other local housing associations. This was to have been interest-free, but it was discovered that this would have been illegal so interest had to be charged at 4%. The loan is for 25 years: redemption begins after the first 10 years, and takes place in 15 equal parts during the remaining 15 years.

The plot has now been planted over. It is considered undesirable to build housing upon it again and it will probably become part of a larger open space stretching northwards. It is still owned by the housing association.

At the same time as block S8 was demolished, it was decided to improve the immediate surroundings in the estate. The most radical change will be the demolition of nine houses in order to break up the uniformity. That will mean that 109 of the 418 dwellings have been removed. (see Fig. 10.2) The remaining dwellings are to be improved.

The infrastructure will also be improved. Parts of the public open space are being handed over to the residents to use as private gardens. In this way, the municipality is relieved of some of its maintenance responsibilities, and the area will become more varied.

The cost of these building works (apart from demolishing block S8 but including demolishing the nine houses, which cost 200,000 Hfl) is 10 million Hfl, to be covered by 4 million Hfl from housing improvement subsidies (see Ch. 7.3), 1.5 million Hfl from rent increases and 4.5 million Hfl from the Leeuwarderadeel housing association out of reserves. The cost of the improvements (excluding demolition) amounts to 28,670 Hfl per remaining low-rise dwelling and 33,520 Hfl per remaining high-rise dwelling.

Attention is also to be paid to the community structure. Suggestions have been made for an apprenticeship scheme to tackle high unemployment in the neighbourhood, for handing maintenance over to the residents, and for tackling vandalism. The name will be changed from De Meenthe so that the neighbourhood can start afresh.

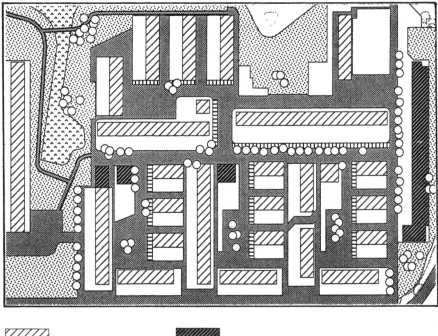

 Dwellings which remain   Dwellings which have been demolished

Figure 10.2 De Meenthe, Leeuwarden.

In the past few years, concrete actions have already been taken. A second resident caretaker was appointed in 1988, the drains have been investigated since there was nuisance from smells in 1989, play facilities have been increased and measures have been taken to slow down traffic and to introduce lighting for footways.

Conclusions

The kind of deterioration experienced at De Meenthe is not uncommon in post-war housing estates, and the way in which it has been tackled in Leeuwarden is being seen as an example for others to follow. The action was drastic (including demolishing 100 flats and 9 houses), but it appears not only to have halted the downward spiral, but also to have changed the whole climate. Whereas between 1985 and 1988 the vacancy rate was around 60% in high-rise dwellings and 30% in low-rise, this fell to zero after the improvements.

The remarkable improvements have been brought about not only by the building works, traffic schemes and environmental improvements, but also by the greater involvement of the residents. By means of various projects, the community structure has been strengthened. Those involved with De Meenthe know, however, that continued attention is necessary to prevent the neighbourhood going downhill again.

10.3 Het Veemarktterrein, Amsterdam: The re-use of urban land for new industrial premises

Context of the development

Amsterdam is the capital of the Netherlands. It is part of the Randstad, and there are 700,000 inhabitants within the municipal boundaries (see general location map). The old cattle market (Veemarktterrein) was in the south-west corner of the Eastern Port Area (Oosterlijk Havengebied, see Fig. 10.3). It was bounded on the north by the Cruquiusweg, on the east by the Veelaan, on the south by the Nieuwe Vaart, and on the west by the Schutsluis. It is rectangular in shape, 335m by 120m (an area of 4.02 ha). It was officially designated for trade and industry, but after the departure of the cattle market it had been used illegally as a caravan settlement.

On 15 August 1974 it was decided to move the cattle market and a committee was set up to advise on a new use for the area. This committee was composed of representatives of the municipality and business, and it acquired the name of OBOS (Ontwikkeling Bedrijfsruimte voor Opvang uit Stadsvernieuwingsgebieden).

This committee started from the principle that the area should be made suitable for accommodating firms in urban renewal areas, firms which because of their size or the nuisance they caused could not remain in their old locations. Some of them were so dependent on suppliers and customers in the older central districts that they might not survive a move to a site on the edge of or outside the city.

Another principle was that the land should be supplied cheaply, or else the rental levels would be too high for the target group. Low rents would also make the development suitable for new firms, for these cannot afford land prices on most industrial estates. The committee also decided that the space should be offered for rent in complexes (*bedrijfsverzamelgebouwen*) or in the existing buildings along the edge of the site.

Finally, it decided that because this site was the only one still available on which firms moving from urban renewal areas could be accommodated, this target group should get priority when the space was being let.

These principles were adopted because urban renewal in the nineteenth century districts was drastically reducing the amount of industrial space there. Small tradesmen and manufacturers were being badly hit. In a muni-

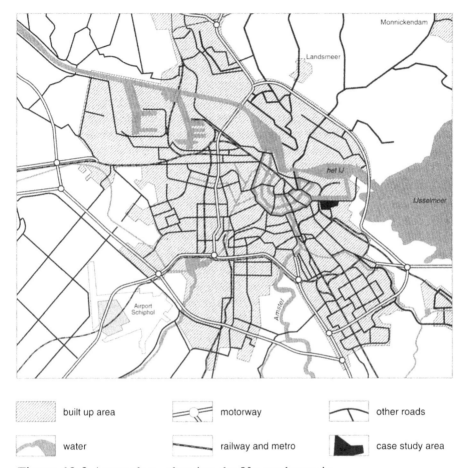

built up area   motorway   other roads

water   railway and metro   case study area

Figure 10.3 Amsterdam, showing the Veemarktterrein.

cipal report about employment (*Nota werkgelegenheid* 1979), it was proposed that industrial land at favourable prices should be provided near to those older districts and that rental complexes should be built on such land. Further, prices for industrial land provided by the municipality on the edge of the city should be more varied.

The land-use plan (*Veemarkt*) for this area was adopted by the municipal council on 19 October 1983. It was approved by the province on 11 December 1984. Two appeals against it were made to the Crown, but these were declared inadmissible. The plan was finalized on 7 January 1986.

The content of the plan

The plan proposed new building on the centre of the site, with $10,000\,\mathrm{m}^2$ industrial space on the ground floor and $2,400\,\mathrm{m}^2$ above that; renovating the existing buildings on the edge of the site to provide 2,900 m², with the possibility of adding another 2,000 m²; or alternatively, if conditions required it, replacing the existing buildings with new single-storey premises.

Environmental aspects had to be considered as the site is adjacent to a residential area, the Indische neighbourhood. The written statement for the plan contains a list of industries, classified by the degree of potential environmental damage. On the site, firms in categories I to IV are permitted, and there is the possibility of granting exemptions for firms in category V which could cause more environmental damage.

There must be one parking space for every 100 m² of industrial floorspace. If that proves to be too little, more spaces can be provided on the land designated for traffic circulation, to a maximum of 1.4 spaces per 100 m². The site is to be connected via the Cruquiusweg (which does not need widening) to the ring road (Stadsring). New buildings must be designed to take account of the character of the buildings that are to be retained.

Subsidies were made available by the municipality (department of economic affairs) for building development on condition that at least 75% of the new premises are occupied by firms moving out of the urban renewal areas in the older central districts. About 4 million Hfl was granted for the first three phases. To prevent speculation a condition has been imposed that lays down a maximum price if the developer sells the project within 10 years. After 10 years, the property can be sold freely. The subsidy was financed out of the urban renewal fund made available to the municipality by the central government (see Ch. 3.3). The same fund can also be used to give temporary income support (e.g. towards paying the higher rents) to firms displaced from urban renewal areas.

Plan implementation

The municipality and the OBOS began discussions with MBO, a developer specializing in shops, industrial premises and offices. The new development was given the name Bedrijvencentrum Oost (Business Centre East).

The site was to be delivered cleared, except for four large houses, which would be retained, restored and used for offices. Further, the site should be ready for building: in practice the foundations of the previous buildings were encountered 2m below the ground and this necessitated some last minute adjustments during the building. When ownership was transferred, awareness of soil contamination was not high (see Ch. 3.3): the developer was later relieved to discover that there was hardly any contaminated soil.

The developer acquired a ground lease on the site for 55 years. The land development process (see Ch. 4.1) was undertaken by the municipal land department and this set the value at 270 Hfl per square metre. Considering the site's central location, this was not high, but it was high considering the rents the displaced firms had been used to paying. The price of the ground lease was set, however, not per square metre land but per square metre lettable floor space. Setting disposal prices in this way gives the land developer, i.e. in this case the municipality, a financial interest in realizing high densities. Instead of paying an annual ground rent for 55 years, the developer paid a premium, amounting to 255 Hfl per square metre of lettable floorspace.

It was decided to build in four phases. The first began in 1984, the second and the third in 1989, and the fourth in mid-1991. The design of the last phase has not been without problems. It occupies a part of the site with a long frontage to the road. The previous building had presented a continuous high frontage, and the municipality had wanted to preserve that effect, as well as the building itself. The developer was convinced that preserving the building was not commercially viable but was prepared to put up a two-storey building, although one-storey would have been better commercially, in order to preserve the urban effect. These changes necessitated a modification to the land-use plan.

The first phase contains about $6,000 \, m^2$ of which $1,200 \, m^2$ is on the first floor. The units have ground floor areas of $10 \times 23 \, m$ ($230 \, m^2$) and 13.5×5.5 ($75 \, m^2$). Some units have a first floor as well, and it had been intended to let these separately as office space. But this was not a success, and in most cases first-floor space was let to the tenants who had rented the ground floor units.

The second and third phases together contain about $10,000 \, m^2$, of which $4,200 \, m^2$ is on the first floor. In these phases, the ground and first floor were let as one unit. The units have ground areas of $23.6 \times 10.6 \, m$ ($25 \, m^2$) and $14.2 \times 5.2 \, m$ ($74 \, m^2$). It will be seen that most of the units are small, and this is also the policy for the fourth and final stage.

The units are intended for small firms in trade and manufacturing. The conditions formally imposed to exclude firms that would cause a high level of nuisance have not in practice had to be invoked. This is because small firms usually cannot undertake the processes that cause most nuisance. Moreover when firms apply to the landlord for a unit they are selected on environmental grounds, so that management problems do not arise.

Letting is done by the development company itself, but after consultations with the municipality. This serves to ensure that the terms of the subsidy are met (i.e. 75% of the units let to firms coming from renewal areas), and that priority is given to firms that must move urgently so as not to delay renewal projects. It is not necessary to advertise the units. Potential tenants often hear about them from the municipal department that is trying to acquire their old premises. This might be why 85% (well over the 75% required) of units are occupied by firms from renewal areas.

Annual rents for first-phase units are around 80 Hfl per square metre for the ground floor and 55 Hfl on the first floor. For the second and third phases, these are around 100 Hfl and 65 Hfl, respectively. These levels mean that most firms are paying three times or more their previous rent. And yet the units are very easy to let, there is a long waiting list and no vacancies. Obviously, the location is very attractive and the subsidy keeps the rents from being too high. Compare the Alpha Business Park (Ch. 6.3) where annual rents are 195 Hfl per square metre.

All rents are annually reviewed, and the developer prefers to let vacant units at the same price as comparable units already occupied, although the demand is high enough for higher rents to be charged for re-lets. The developer does this to make it easier to sell the whole complex when it is finished. MBO is a developer, not a property investment company, although when it does sell, it would like to be retained to manage the complex.

Conclusions

In many respects, this can be regarded as a most successful redevelopment. A site with a good location for industry was cleared of a use for which it was no longer appropriate, and now provides premises for firms that had mostly been operating in the older central districts; these firms operate more efficiently and can survive the change to higher rents because these are subsidized. Moreover the firms do not stand in the way of desirable urban renewal.

However, the project does raise questions about the financial involvement of the municipality and about how this is organized. The first point is that the land development is in the hands of the municipality's land department (*grondbedrijf*). This has an interest in disposal prices being as high as the market will bear. It charged 270 Hfl per square metre, which is high in comparison with the prices on most industrial estates (see Ch. 5.3). However, it would not be high for offices, for which this location would have been suitable.

The second point is that another department of the municipality pays a subsidy to the developer to keep rent levels down. The amount of the subsidy is calculated in such a way that the higher the rent charged to the tenants, the lower the subsidy. This department therefore has an interest in the developer charging high rents. The developer, however, has no interest in this: the subsidy is lowered if rents are raised, and higher rents make letting more difficult.

The third point is that the level of subsidy is dependent also on the development costs: the higher the costs, the higher the subsidy. The land department, by setting high disposal prices, pushes up costs, for which another department must compensate with higher subsidies.

The fourth point is that the same department of the municipality that pays the subsidy for the building can pay temporary income subsidies to firms that move from urban renewal areas into more expensive premises, in order to help them adjust to higher costs. So higher rents reduce one subsidy paid by the department, and increase another that it pays.

The final point is that the land department sets disposal prices (the ground lease) as an amount per square metre of lettable floor space. It therefore has a financial interest in high building densities. Moreover, the planning department also wants the site to be used as intensively as possible. It is centrally located and there are many smaller firms to be accommodated. However, the developer has little interest in building more than one storey, for first-floor industrial premises are difficult to let.

The municipality has a great financial involvement in this project. These concluding points show that financial involvement can have unexpected and unwanted effects.

PART IV
Evaluation

CHAPTER 11
Evaluation of the urban land and property markets

Having described the legal and financial framework of the urban land market and the processes that take place within it; explained the prices, quantities, etc. that are the outcome of the processes within that framework; and considered the urban property market in the same way, this concluding chapter aims to evaluate how these markets work.

In the Netherlands it is relatively easy to separate the urban land market from the urban property market because the actors are different and their rôles clearly distinguished. Nevertheless, it makes little sense to evaluate the two markets separately. This is because the main function of the urban land market is to enable the urban property market to work, and so the criteria for judging the urban land market must be derived from how the urban property market works. Hence, this evaluation treats the two markets together.

For this study it was agreed that the functioning of urban land and property markets should be evaluated according to the criteria listed below.
1. There should be a sufficient supply of property. Sufficient relates to need as well as demand, supply should be sufficient for each type of use, including the needs and demands of poorer people and non-commercial activities. Sufficient relates also to location of supply relative to the location of need or demand: the supply should be in the right place.

   Supply should be able to react quickly to changes in need and demand: to some extent, sufficiency can be interpreted in terms of prices. Supply is insufficient if prices are higher than need be; supply is more than sufficient if prices have fallen so low that further production is not profitable. Sufficient must be interpreted in the light of public policies for the location of activities (such as regional policy, physical planning policy and ecological protection). If supply is restricted because of these policies, this does not mean that supply is insufficient. Sufficient must not be interpreted in such a way that land is used wastefully, since land should be seen as a finite resource, and natural areas have an ecological value.
2. It should be possible for public policy to influence not only the amount

and location of the property (the sufficiency of supply) but also its nature and use, for example so that one person does not impose negative external effects on others, and so that unnecessary pollution or use of finite resources is avoided.

3. The private ownership of property should not be too unequally distributed. This is important for financial reasons (property is wealth), and for socio-political reasons (ownership of property can be abused to exercise power over others).

4. Prices for property should not fluctuate greatly. Fluctuating prices cause uncertainties that can destabilize the property market (and possibly other markets as well), and they can cause speculation in property, which interferes with a steady supply.

5. The processes on the property market, and not only the outcomes, should be visible and socially acceptable. All the actors should have the certainty of working within a predictable legal framework; private actors should not be subject to unnecessarily frequent, arbitrary, or unfair intervention and control by public administration (e.g. by unpredictable or unreasonable use of compulsory purchase powers, or actions that unfairly reduce prices or values); and private actors should not be able to make excessive gains on the property market caused by public actions (e.g. development gains caused by granting planning permission).

How does the Dutch property market measure up to the standards listed above? We believe quite well.

The practice whereby building land is supplied by municipalities which see themselves as supplying a public utility as cheaply as possible (see Ch. 4.2) means that the supply is sufficient: the supply of land is not allowed to be a hindrance to the supply of property, unless it is public policy that it be hindered. And the willingness to take advantage of this opportunity by building on that land, coupled with the financial ability to do so, means that the supply of property is sufficient in the terms of criterion 1.

This is a conclusion that might also be significant for other countries. If building land is supplied "on tap" by public authorities that want to encourage development (but at times and in locations determined by public policies) without making profits out of that supply, then property developers will treat building land as just one of the many factors of production. And if the capacity of the construction industry and the supply of development finance are adequate, then buildings will be provided in a way consistent with public policy (criteria 1 and 2) by private developers operating commercially.

Moreover, this is achieved in the Netherlands without the distribution of private ownership of property being regarded as a political issue (criterion 3); without prices fluctuating in a destabilizing way (criterion 4) - except for the house price boom and slump of 1978–82 (see Ch. 9.3); and without processes on the property market being regarded as an economic or political issue (criterion 5).

That rosy picture is not, however, without its thorns. Some of these are discussed below.

1. The price of rented housing is high and rising (see Ch. 9.3). This is not because the price is unnecessarily high: most rented housing provides good value for money and stands on land that has been subsidized. Rather, it is because some people cannot afford the minimum quality of housing below which Dutch society does not want to let people live (a minimum level high in comparison with most other countries). This is to be regarded more as a problem of the relationship between incomes and housing policy than as a malfunctioning of the urban property market.

2. The proportion of households owning their own homes is, compared internationally, quite low (see Ch. 9.1). This does not contribute to a more equal distribution of wealth, nor to greater control over one's own living conditions. It is current public policy to increase owner occupation (see Ch. 2.4). If the government hopes thereby to reduce housing subsidies, it may be disappointed. It is not certain that subsidies paid for a dwelling that is rented are higher than tax relief on mortgage interest if that dwelling is owner occupied (see Ch. 7.3). Ironically, by giving such high mortgage relief (there is no upper limit) the government makes the stimulation of owner occupation both more expensive for the public purse and more necessary in order to equalize the wealth that can be acquired by owner occupation.

3. There is no public discussion about whether land is being used wastefully. This is curious in a country as small and densely populated as the Netherlands, where one would expect intense debate before taking each hectare of land from rural into urban use. Yet the rate of conversion has raised the proportion of the total land area used for urban purposes from 9% in 1970 to 13.7% in 1985. In that period, the area in urban use has increased by 55% while the population has increased by 12%; and the urban area per person has increased from 254m^2 to 354m^2 (see Ch. 1.4). Moreover, it has been estimated (Bethe 1991) that in the period 1990–2000 another 17,300ha will be taken into urban use plus 10,000 ha for infrastructure and industrial estates. This is, in ten years, another 273km^2 of urban land on top of the 5,110km^2 already (1985) in urban use.

The fact that this is not a public issue does not of itself mean that land is being used wastefully. Housing densities, for example, are quite high by international standards. Lindenholt (see Ch. 6.1) can serve as an example of a modern low-rise housing estate. The whole development covers 300 ha, and that includes not only housing and its associated services but also offices, shops, schools, etc. When it is finished, about 6,000 dwellings will have been built there. At 20 dwellings per hectare gross, land is not being squandered.

Where land is (or has been) used wastefully is for industrial estates. In the 1970s, thousands of hectares were zoned for this use (see Ch. 5.2). When it became clear that this was a massive oversupply, some was rezoned for rural uses. But that so much rural land could be zoned speculatively for urban uses illustrates the lack of social and political concern about stewardship over rural land. The attitude is that urban needs always have to be met. This is a clear result of the way municipalities see their rôle as land developers, whereby building land must always be available on tap.

4. Developers of and investors in commercial property complain about the returns on their activities. They complain that rental levels and the capital value of their properties are rising too slowly, making the investment climate in the Netherlands less attractive than that in other countries. The reasons for that price stability are indeed to be found in the way the land and property market functions, apart from the fact of low inflation, which also helps to keep property prices stable.

Two of the main points have been given in Chapter 5.2: the suppliers of land try to prevent any shortage of opportunities for building, property developers take up those possibilities, so there is no shortage of buildings, so prices do not rise. Also all suppliers try to provide good locations, so the geographical variation in prices is small.

A third point is that the supply of land is so predictable. It is supplied in accordance with a land-use plan at times and locations decided by public bodies, not by private developers.

Property developers should be able to make normal profits out of the building development process (see Ch. 4.2). They make extra profits by being able to detect and realize development opportunities that their competitors have not grasped. Analyzed in this way, property developers in the Netherlands are complaining that they cannot make extra profits, because land supply is so predictable that it is difficult to steal a march on competitors; and that sometimes it is difficult even to make normal profits. The reason for this must be that the developers are taking up too many of the building opportunities made available to them. Municipalities might be glutting the market for building land, but it is the property developers that are glutting the market for commercial properties.

Property investors make normal profits if they acquire property at a price that reflects the expected returns from that property and if everyone shares those expectations. They make extra profits if the actual returns are higher than the market had expected. Analyzed in this way, property owners are complaining that they cannot make extra profits because everyone expects the same returns, as returns are so predictable; and that sometimes they cannot even make normal profits because the expected returns are too low relative to the acquisition price. The reason must be that investors are competing against each other to acquire property, which drives up the prices. This makes it profitable for property developers to supply more property, which drives down the rents.

Seen in this way, one is not sympathetic towards the property developers and investors: if there is an oversupply, that is for the industry itself to resolve. Moreover, if oversupply results in low prices, this is advantageous for the consumers of urban property: and the urban land and property market exists for the consumer, not for the producer.

However, this situation does illustrate the general rule that the commercial property market cannot regulate itself. It would benefit from regulation imposed on it, and it is only a public body can do this. The questions raised by the property industry lead to consideration of the following points, which are increasingly being seen as a malfunction of the urban land and property market.

5. Certain types of property are spatially too dispersed and should be more concentrated. We have seen how small the geographical variation in property prices is (see Ch. 9.3) and this has been explained by the wish of all municipalities to attract development (Ch. 5.2) and the lack of action by provincial or national government to restrict it. In other words, there has been no policy for concentrating development.

One of the results has already been mentioned: more land is converted from rural to urban use than is necessary. Secondly, certain services that need to be clustered in order to function well are too dispersed. This applies in particular to firms providing services nationally and internationally. The Netherlands has, for example, no office centre that can compete internationally. Offices are spread between Amsterdam, Rotterdam and The Hague, with Utrecht growing rapidly, and there are also suburban office clusters around the four big cities and along some motorways. Thirdly, journeys to work are too dispersed, which makes it difficult to accommodate them on public transport, placing an undue reliance on private cars, with all the consequences for road construction, car parks, petrol consumption, air pollution and road accidents.

For these reasons, attention is increasingly being paid to introducing a concentration policy. The beginnings of this were described in Chapter 2.1. This would have to be implemented by central government and would be fiercely contested by many (probably most) of the 700 municipalities.

What would be the effects on the property market? And would it meet the complaints of the property developers and investors? The effects would be an increase in property prices in the chosen "concentration areas" and a decrease (or a stabilization) elsewhere. Those owning existing properties in the concentration areas would benefit (capital growth), those owning existing property elsewhere would not. Rents and capital values for properties yet to be developed in the concentration areas would be higher than without the policy. If the municipalities as land suppliers adjusted disposal prices accordingly, the municipalities would benefit, not the developers. Outside the concentration areas, land values would fall, a disadvantage to those municipalities. It is not certain whether prices in the concentration areas would continue to rise after the introduction of the concentration policy: hence it is not certain that property owners would benefit from a continuing income and capital growth.

In short, some kind of concentration policy would probably be in the general interest. It would not, however, be in the interest of all municipalities and their inhabitants, nor in the interest of all property developers and investors. Moreover, the benefit to property developers and investors might be a temporary effect caused by introducing the policy, rather than a permanent effect of the continued application of the policy.

REFERENCES

Aalders, M. 1991. Netherlands, organizational structure. In *European environmental yearbook 1991*, A. Cutrera (ed.), 342–4. London: DocTer.

Adriaansens, C. A. & A. Ch. Fortgens 1986. *Volkshuisvestingsrecht* (3rd edn). Deventer: Kluwer.

Arthur Andersen & Co. 1991. *Investment in Netherlands real estate*. Amsterdam: Arthur Andersen.

Atzema, O. 1991. *Stad uit, stad in*. Utrecht: University of Utrecht (RUU).

Bak, I 1986. Winkelcentra in Nederland. *VastGoedMarkt*, 11 April.

Bangma, A. & J-P. Kruimel 1990. *Onroerend-goedbelasting: hoe en wat*. Kluwer belastingwijzers 18. Deventer: Kluwer.

Beckers, Th. & H. van der Poel 1990. *Vrijetijd tussen vorming en vermaak*. Leiden/ Antwerpen: Stenfert Kroese Uitgevers.

Belastingpraktijkboek voor de belegger 1986/1987. *Onroerend goed*. Deventer: Kluwer.

Beld, J. L. M. van den 1991. Het ABC van Alders: werken bij het station. *Financieel Economisch Management* **4**.

Belfield & Everest (eds) 1988. *Spon's international construction costs handbook*. London: Spon.

Bethe, F. H. 1991. *Regionale grondbalansen, een verkenning van de behoefte aan en de beschikbaarheid van grond tot en met het jaar 2000*. 's-Gravenhage: Landbouw Economisch Instituut.

Blankenstein-Bouwmeester, A. & P. Lukkes 1984. *Institutionele beleggers op de markt voor onroerend goed*. Sociaal-geografische Reeks nr. 29. Groningen: University of Groningen.

Bovy, C. & P. Ekkers 1990. *Kavels en woningen in Nederland II*. Den Haag: Min. VROM.

Bovy, P. H. L. & D. N. den Adel 1988. De groei in het forensisme (1) en (2). *Verkeerskunde* **40**, 385–9

Breevast 1984 *Vastgoed en de Kapitaalmarkt*. Utrecht. 435–7

Broek, P. J. van den 1988. *Overheid en grondprijzen, het beleid sinds de jaren vijftig*. Amsterdam: Stichting voor Economisch Onderzoek, Amsterdam.

Brouwer, J. 1990. *Ruimte voor investeringen*. Delft: Waltman.

Bruil, D. W., A. Faludi, H. J. Gastkemper 1987. *Dutch land-use laws*. Amsterdam: Werkstukken Planologisch & Demografisch Instituut, University of Amsterdam.

REFERENCES

Brussaard, W. 1987. *The rules of physical planning*. Den Haag: Ministry of Housing, Physical Planning & Environment.

Buijs, C. C. M. & H. I. E. Dijkhuis-Potgieser 1989. *Ontwikkelingen op de woningmarkt 1989*. Den Haag: Min. VROM.

Centraal Bureau van de Statistiek (CBS: Central Bureau of Statistics). Many different statistics, e.g. Statistisch Zakboek, Maandstatistiek Bouwnijverheid, Maandstatistiek voor de Prijzen, Statistiek van het Bodemgebruik. Den Haag.

Centraal Planbureau (CPB) 1990. *Economisch Beeld, 1991*. Den Haag: CPB.

Dekker, A. 1991. Town and country planning. In *European environmental yearbook 1991*, A. Cutrera (ed.). London: DocTer.

Dijkstra, O. A., E. Koperdraat, W. de Weijer 1986. *Inleiding Wet op de Ruimtelijke Ordening en Woningwet*. Alphen aan de Rijn/'s-Gravenhage: Samsom/ VUGA.

Economisch Instituut voor de Bouwnijverheid (EIB, Conijn et al.) 1980. *De financiering van investeringen in bouwwerken*. Amsterdam: EIB

Economisch Instituut voor de Bouwnijverheid (EIB) 1988. *Bouwmarkt en midden- en kleinbedrijf in 1995*. Amsterdam: EIB/Min. VROM/Ministerie van Economische Zaken.

Economisch Instituut voor de Bouwnijverheid (EIB) 1991. *Ontwikkelingen in de bouwnijverheid tot en met 1996*. Amsterdam: EIB/AVBB.

Fenten, G. 1985. Leegstaande kantoren: wat gaan we ermee doen? Final-year dissertation, Vakgroep Planologie, University of Nijmegen.

Gemeente Amsterdam 1990. *Gemeentelijk grondbeleid, modernisering erfpacht stelsel*. Amsterdam.

Gemeente 's-Gravenhage 1983. Discussienota grondbeleid. 's-Gravenhage.

Geus, J-W. de 1990. Kantorenlokaties en rendement. In Amsterdam, Den Haag, Rotterdam en Utrecht. Final-year dissertation, Vakgroep Planologie, University of Nijmegen.

Giebels, R., C. C. Koopmans, F. Moolhuizen 1985. *De risico's voor gemeenten op de markt voor bouwrijpe grond*. Amsterdam: Stichting voor Economisch Onderzoek.

Gilhuis, P. 1991. Netherlands, legislation. In *European environmental yearbook 1991*, A Cutrera (ed.), 830–31. London: DocTer.

Goede, B. de, & B. Troostwijk 1956. Het gebruik van burgerrechtelijke vormen bij de behartiging van openbare belangen. In *Pre-adviezen voor de algemene vergadering van de Vereniging van Administratief Recht*. Haarlem: Tjeenk Willink.

Graaf, B. de 1984. De exploitatie van grond. Paper given at conference Inleiding in het gemeentelijk grondkostenbeleid. Amersfoort.

Groot, E. de, H. Kerkhof, L. Veening 1988. *Land-use changes in the Netherlands*, Mededelingen Vakgroep Cultuurtechniek no. 124. Wageningen: Agricultural

REFERENCES

University of Wageningen.

Heide, H. ter, 1991. Planning in the Netherlands, Paper to the training course "Planning in the Netherlands" held by the Royal Town Planning Institute, London.
Hereijgers, A., P. Roelofs, D. Schuiling 1989. *5 PPS-projecten voor stedelijke vernieuwing.* 's-Gravenhage: Min. VROM.
Hogenbirk, J. & A. Loggers 1988. Dynamiek van de commerciële vastgoedmarkt 1983–1988. Final-year dissertation, Fakulteit der Ruimtelijke Wetenschappen, University of Utrecht.
Hollenberg, A. 1978. *Grondpolitiek in Nederland.* Doetinchem: Misset.

Ike, P., H. Voogd, K. van Zwieten 1984. Planning voor bedrijfsterrein; gokken met ruimte en overheidsgeld. *Economisch Statistiche Berichten,* 494–7.

Jansen, G. R. M. & T. van Vuren 1987. Het externe verkeer van de grote steden. *Verkeerskunde* **38**(10).
Jansen, J., J. Bloemberg, D. Hardy et al. 1985. *De compacte stad gewogen.* Studierapporten Rijksplanologische Dienst 27. 's-Gravenhage: Min. VROM.
Janssen, J. E. 1992. *Determinanten van prijzen voor bestaande koopwoningen.* Nijmegen: Vakgroep Planologie, University of Nijmegen.
Jones Lang Wootton 1990 *Quarterly investment report; the European property market.* Amsterdam: Jones Lang Wootton.
Jonge, J. de 1984. *Gemeentelijke gronduitgifte.* Dewenter: Kluwer.

Kadaster en Openbare Register (KADOR) 1991. Dienst Juridische Zaken en Vastgoedinformatie. Onderzoek prijsontwikkeling particuliere woningen (periode november 1977–februari 1991). Apeldoorn: KADOR.
Koopmans, L. & A. H. E. M. Wellink 1987. *Overheidsfinanciën.* Leiden: Stenfert Kroese.
Kruijt, B. & J. E. Janssen 1991. Een marktstemmingsindex voor vastgoed.*Econo misch Statistische Berichten,* 23 January.
Kruijt, B., D. B. Needham, T. Spit 1990. *Economische grondslagen van grondbeleid.* Amsterdam: Stichting voor Beleggings en Vastgoedkunde/University of Amsterdam.

Lie, R. & A. Bongenaar 1990. *Toplocaties: produktiemilieu én investeringsmilieu.* Amsterdam: Stichting voor Beleggings en Vastgoedkunde/University of Amsterdam.
Lijphart, A. 1968. *The politics of accommodation.* Berkeley: University of California Press.
Lukkes, P., A. J. Krist, P. J. M. van Steen 1987. *Kantorenmarkt, investeren en ruimte.* Zeist: Vonk Uitgevers.

Ministerie van Volkshuisvesting (Min. VROM). *Ruimtelijke ordening en milieu beheer.*

219

REFERENCES

Min. VROM 1988. *Rijkshuisvestingsplan: de vierkante meter van morgen.* 's-Gravenhage: Min. VROM.

Min. VROM 1989a. *Meerjarenplan Stadsvernieuwing 1990-1994.* 's-Gravenhage: SDU uitgeverij.

Min. VROM 1989b. *Nationaal Mileubeleidsplan 1989.* 's-Gravenhage: SDU uitgeverij.

Min. VROM 1989c. *Nota Volkshuisvesting in de jaren negentig.* 's-Gravenhage: SDU uitgeverij.

Min. VROM 1989d. Directoraat-Generaal van de Volkshuisvesting. *Volkshuisvest ingsinstrumenten 1990.* 's-Gravenhage: Min. VROM.

Min. VROM 1990a. *Nationaal Milieubeleidsplan – plus.* 's-Gravenhage: Min. VROM.

Min. VROM 1990b. *Vierde nota over de ruimtelijke ordening – extra; deel 1: ontwerp-planologische kernbeslissing* (Vinex). 's-Gravenhage: SDU uitgeverij.

Min. VROM 1990c. *Volkshuisvesting in cijfers 1990.* Zoetermeer: Min. VROM.

Min. VROM 1990d. *Leegstand en hergebruik, een nieuwe markt voor de bouw.* 's-Gravenhage: Min. VROM

Nationale Woningraad (NWR), R. J. H. Docter et al. 1990. Monumentenzorg en volkshuisvesting. Woningraad extra 51 Almere: NWR.

Nederlands Instituut voor Ruimtelijke Ordening en Volkshuisvesting (NIROV) 1985. *Onderzoek naar de oorzaken en achtergronden van de prijsontwikkeling van woningkavels.* 's-Gravenhage: NIROV.

Needham, D. B. 1991. The land and property markets in the Netherlands (paper). University of Nijmegen.

Needham, D. B. 1992. A theory of land-prices when land is supplied publicly. *Urban Studies* **29**, 669–89.

Needham, D. B. & R. Beljaars 1988. Financiële aspecten van de volkshuisvesting. In Needham, B., *Syllabus, bouwen en wonen.* Nijmegen: Vakgroep Planologie, University of Nijmegen.

Needham, D. B. & B. Kruijt 1992. The Netherlands. In *Industrial property markets in western Europe*, B. Wood & R. H. Williams (eds). London: Spon.

Neuerburg, E. N. & P. Verfaille 1985. *Schets van het Nederlands milieuhygiëne recht.* Alphen aan den Rijn/'s-Gravenhage: Samsom/VUGA.

Nozeman, E. F. 1988 Groeikernen. In *Successen en mislukkingen in de Nederlandse Ruimtelijke Ordening*, C. Bos et al. (eds). Amsterdam: Planologisch en Demografisch Instituut, University of Amsterdam.

Nozeman, E. F. 1990. Dutch new towns: triumph or disaster? *Tijdschrift voor Economische en Sociale Geografie* **81**, 149–55.

OECD. Annual National Accounts, Main Aggregates 1960–1986 and 1987–1989, (on diskettes). Paris: OECD

Olden, H. 1984. *Bedrijfsterreinen in Nederland*, Den Haag: Rijksplanologische Dienst.

OTB, Enquete institutionele beleggers. *Tweede Kamer*, vergaderjaar 1987–1988, **19623** nr. 33, p. 71.

REFERENCES

Prins, J. H. A. A. 1985. *De gemeentelijke grondexploitatie*. 's-Gravenhage: Vereniging van Nederlandse Gemeenten.

Provincie Gelderland 1983. *Tussenrapportage over de financiële problematiek bij de exploitatie van bestemmingsplannen*. Arnhem: Provincie Gelderland.

Research Institute for the Built Environment (RIGO) 1987. Documentatiebundel OECD-"position papers"-urban land market, deel 1 en 2. Amsterdam: RIGO.

RIGO 1989. *Het grondprijsbeleid voor de woningbouw sinds 1900*. 's-Gravenhage: Min. VROM.

Rooy de, E. Elbers 1989. *Woningbehoefte-informatie voor marktpartijen*. 's-Gravenhage: VROM, DGVH.

Rotterdam, Municipality of, 1959. *Municipal real estate policy in the Netherlands*. Rotterdam: Gemeente.

Rouwenhorst, D. 1991. De industriële onroerend goedmarkt. Final-year dissertation, Vakgroep Planologie, University of Nijmegen.

Schuiling, D. 1987. Urban renewal: The Netherlands. In *European environmental yearbook 1987*, A. Cutrera (ed.), 600–603. London: DocTer.

Smit, W. A. P. 1989. *Huurrecht bedrijfsruimten*. Deventer: Kluwer.

SOAB 1987. *Plan van aanpak – De Meenthe Leeuwarden*. Breda: SOAB.

Stichting voor Economisch Onderzoek (SEO) 1980 *Onderzoek naar gemeentelijk grondprijsbeleid*. Amsterdam: SEO.

Sociaal en Cultureel Planbureau (SCP) 1982). *Sociaal en Cultureel Rapport 1982*. 's-Gravenhage: Staatsuitg everij.

SCP 1990. *Sociaal en Cultureel Rapport 1990*. Rijswijk: SCP.

Spit, T. 1987. Town and country planning. In *European environmental yearbook 1987*, A. Cutrera (ed.), 554–60. London: DocTer.

TAUW Infra Consult 1990a. *Kavelopbrengsten 1982 t/m 1988, ongewogen*. Deventer: TAUW.

TAUW Infra Consult 1990b. *Ontwikkelingen kavelprijzen nieuwe uitleg 1982–1988*. Deventer: TAUW.

Technische Hogeschool Delft 1981. Collegedictaat B062A03 *Werkgelegenheid*. Delft: TH.

Terpstra, P. R. A. 1991. Zwakke plek in PPS. *Stedebouw en Volkshuisvesting* **72** (April), 39–42.

Thomas, D. et al. 1983. *Flexibility and commitment in planning*. The Hague: Martinus Nijhoff.

Tijdschrift Milieu & Recht. Bodembescherming en sanering (themanummer) (December) 1990/12. Zwolle: Tjeenk Willink.

Uitterhoeve, W. (ed.) 1990, *De staat van Nederland*. Nijmegen: SUN/RUU/ITS.

VastGoedMarkt. From **11**/3 (November) 1989. Amsterdam: 1989-90–91.

Verhagen, J. 1989. Flexibiliteit in de gemeentelijke ruimtelijke-ordenings-praktijk. In *Ruimtelijk handelen*, N. Muller & B. Needham (eds). Zeist: Kerckebosch.

REFERENCES

Vocht, G. de, 1990. Stedelijke vernieuwing: vernieuwing van steden? In *Rooilijn* **90**/9, pp. 252-7. Amsterdam: PDI, University of Amsterdam.

Voorlopige Raad voor vastgoedinformatie (Ravi) 1988. *Vastgoedinformatie.* Apeldoorn/Gouda: Ravi en Van Berkel & Partners.

VROM 1990. *Volkshuisvestingsinstrumenten 1991.* 's-Gravenhage: SDU Uitgeverij.

VROM, DGVH 1991. *Bewoners nieuwe woningen 1990.* 's-Gravenhage: SDU Uitgeverij.

Waard, P. de, 1991. Bankier en advocaat snakt naar duur kantoor in Zuid. *De Volkskrant* (VK), **6**-5.

Wetenschappelijke Raad voor het Regeringsbeleid (WRR) 1990. *Van de stad en de rand.* Rapporten aan de Regering, nr. 37/1990. 's-Gravenhage: SDU uitgeverij.

Willems-Schreuder, J. 1989 *Bewoners nieuwe woningen 1987.* Zoetermeer: Min. VROM/BSM.

Woonbond, 1991. *Bevries de huren.* Amsterdam.

Zadelhoff Research 1990. *Visie achter de feiten 1989–90.* Amsterdam: Zadelhoff Makelaars.

INDEX OF ENGLISH TERMS

INDEX OF DUTCH TERMS

Milton Keynes UK
Ingram Content Group UK Ltd.
UKHW031148141024
449569UK00024B/983